Current Topics in Microbiology and Immunology

229

Editors

Springer
Berlin
Heidelberg
New York
Barcelona
Budapest
Hong Kong
London
Milan
Paris
Santa Clara
Singapore
Tokyo

Somatic Diversification of Immune Responses

Edited by G. Kelsoe and M.F. Flajnik

With 38 Figures

Springer

Professor Garnett KELSOE, D. Sc.
University of Maryland School of Medicine
Department of Microbiology and Immunology
655 W. Baltimore Street
Baltimore, MD 21201
USA

Professor Martin F. FLAJNIK, Ph.D.
University of Miami School of Medicine
Department of Microbiology and Immunology
1600 NW 10th Avenue
Miami, FL 33136
USA

Cover Illustration: Figure and text are from Weigert MG, Cesari IM, Yonkovich SJ, Cohn M (1970) Variability in the lambda light chain sequences of mouse antibody. Nature 228: 1045–1047.

Cover Design: design & production GmbH, Heidelberg

ISSN 0070-217X

ISBN-13: 978-3-642-71986-8 e-ISBN-13: 978-3-642-71984-4
DOI: 10.1007/978-3-642-71984-4

Library of Congress Catalog Card Number 15-12910

Softcover reprint of the hardcover 1st edition 1998

Typesetting: Scientific Publishing Services (P) Ltd, Madras

SPIN: 10565858 27/3020 – 5 4 3 2 1 0 – Printed on acid-free paper

Preface

Discovery of the mechanism for V(D)J hypermutation remains a basic goal of immunology despite the best efforts of many laboratories. The existence of catalyzed, site-specific mutation and its exploitation for the somatic evolution of lymphocytes are remarkable adaptations, yet since the discovery of hypermutation in 1970 (see cover), much hard work has generated little. Indeed, our knowledge of what is probably absolutely required for the mutator's action can be succinctly expressed: *Ig gene enhancers*. Table 1 of Winter et al.'s chapter puts into a historical perspective how our notions of the mutator have changed over the years.

Despite these modest gains, most of us feel that this is the best of times. Our work has not only shown us what the mutator is not, it has also, like an artist's preliminary sketch, defined the questions and experiments we must face without diminishing the potential for new biology. In short, it is great fun to toil against a significant and enigmatic problem.

This volume illustrates the great potential that remains in the search for the mutator. The models for V(D)J hypermutation presented here are rich and varied, precisely because we have so much yet to learn. As you peruse this work, ask yourself the following: (1) What roles do transcriptional regulatory regions play in V(D)J hypermutation and how, when, and where do they act? (2) Are mutation and gene conversion distinct mechanisms and how have they evolved? (3) Is the germinal center necessary for Ig hypermutation and selection or only one or neither? (4) What experimental models should be used to study hypermutation? Can we develop systems to identify B cell-specific mutation factors as powerful as those used to identify *Rag*-1 or *Rag*-2? (5) Finally, what are the evolutionary origins of V(D)J hypermutation and how was the mechanism co-opted by the immune system?

One of us (MFF), hopes for the Ehrlich Prize and bodily transportation up to heaven with the resolution of these fundamental questions; the other (GK) would gladly take the money and remain on earth. But until that time, we both remain excited, optimistic, and ever enthusiastic that the mutator will be soon revealed.

List of Contents

List of Contributors

(Their addresses can be found at the beginning of their respective chapters.)

The Role of Promoter–Intron Interactions in Directing Hypermutation

D.B. WINTER[1], N. SATTAR[2], and P.J. GEARHART[1]

1 Introduction

Evolution is driven by a revolving process of mutation and natural selection during which the fittest individuals survive under harsh selective pressures in their environment. In a molecular recapitulation of this process, antibodies undergo a somatic evolution after antigen stimulation, resulting in a more protective defense of the host against the environment. Immunoglobulin genes undergo apparently random hypermutation of their rearranged variable regions followed by selection for the fittest (i.e., highest affinity) of the mutated antibodies. The first evidence of hypermutation and selection of immunoglobulin genes was the observation of increased mutations in the sequences of variable regions of heavy chains encoding anti-phosphorylcholine antibodies bearing IgG isotypes (GEARHART et al. 1981) and in their associated V_κ genes (SELSING and STORB 1981). Over the ensuing 16 years, we have delineated the pathway that B cells undergo to acquire somatic mutations in their rearranged immunoglobulin genes, in regards to the timing, location, and some extracellular requirements (GEARHART 1993). The advent of transgenic and polymerase chain reaction (PCR) technology brought especially powerful tools to cut and paste the immunoglobulin loci together to establish the minimal sequences required for targeting hypermutation, and gene targeting allows the targeting of proteins that may play a role in signaling the cell to begin mutating. Flow-activated cell sorting permits the capture of small populations of

[1]Laboratory of Molecular Genetics, Box 01, Gerontology Research Center, NIA, NIH, 4940 Eastern Avenue, Baltimore, MD 21224, USA
[2]National Institute of Transplantation, 2200 West Third Street, Los Angeles, CA 90057, USA

hypermutating B cells, from which DNA or RNA may then be amplified by PCR, cloned, and examined. Yet for all our studies, the molecular mechanism actually involved in creating the mutations remains a mystery. Furthermore, in the last few years much of what we thought we knew about hypermutation has come under fire. Table 1 presents elements commonly thought to be required for hypermutation, and beside them are elements (many are the same) which have been shown in the last few years *not* to be required. The contradictory nature of the recent data is both confounding and exciting. Although it may temporarily confuse us, when we resolve the contradictions, we will have gained much greater insight of the mechanism itself.

2 Cis-DNA Sequences Required for Targeting Hypermutation

One of the central questions about the mechanism is how it carries out its function with such exquisite specificity. The hypermutation mechanism precisely targets rearranged immunoglobulin genes. This requires local cis-DNA sequences to guide the mutation machinery to the proper site. Work from several laboratories has linked the frequency of hypermutation with both the J-C intronic enhancer/matrix attachment region (E/MAR) and 3′ enhancer (3′ E) several kilobases downstream

Table 1. Requirements for hypermutation

Required for hypermutation	Not required for hypermutation
B Cells	B Cells (Zheng et al., 1994)
Immunoglobulin receptor	V(D)J gene, Ig promoter (Betz et al., 1994; Yélamos et al., 1995)
Antigen stimulation	Antigen stimulation (Reynaud et al., 1995)
T Cell help	αβ T cells (Dianda et al., 1996)
Germinal centers	Germinal centers, lymphotoxin-α (Matsumoto et al., 1996)
Transcription	
Rearranged V(D)J	IgD, TdT, CD23, CD30, IL-4 (Texido et al., 1996)
Enhancers	CD40 (Wykes et al., 1997)
	DNA helicase (Green and Sack, 1997)
	Nucleotide excision repair (Wagner et al., 1996)
	Mismatch repair (Winter and Gearhart, unpublished)

IL-4, interleukin-4.

of the constant gene (MEYER and NEUBERGER 1989; HACKETT et al. 1990; SHARPE et al. 1991; GIUSTI AND MANSER 1993; SOHN et al. 1993; BETZ et al. 1994). Other groups have demonstrated that most mutations occur downstream of the promoter, suggesting that the promoter is also important for hypermutation (LEBECQUE and GEARHART 1990). These data are consistent with the notion that hypermutation is linked to transcription. Paradoxically, hypermutation theoretically occurs only in B cells residing in the dark zones of germinal centers, where there are only low levels both of immunoglobulin mRNA and cytoplasmic Ig proteins (CLOSE et al. 1990; LIU et al. 1992). The role of the enhancers and transcription, therefore, may be to open up the gene and make it accessible to the hypermutation machinery.

This leaves unanswered the question of the cis-DNA sequences which specifically target the mechanism to the 5′ end of the gene. While the epicenter of mutation is the rearranged V(D)J gene, the targeted area reaches into both the 5′ and 3′ non-coding regions. On the 5′ side, the mutations end abruptly, and the distance appears to be linked to the distance between the rearranged gene and the promoter (WEBER et al. 1994). On the 3′ side, the mutations trail off after about 1000 bases in the J-C intron (GEARHART and BOGENHAGEN 1983). In order to determine if DNA sequences in or around the epicenter might act as a homing signal for the mutation mechanism, YÉLAMOS et al. (1995) removed most of the coding region of a V_κOx-1 transgene and substituted it with similar lengths of three other genes. Transgenic mice bearing these constructs were able to mutate them at a frequency similar to that of the wildtype construct, suggesting that the coding region itself is not required for hypermutation. Furthermore, BETZ et al. (1994) replaced the promoter with the human β-globin promoter and still retained hypermutation in the coding region, although at a slightly lower level than the wildtype. These data are consistent with the idea that an immunoglobulin promoter itself is not essential for targeting hypermutation.

3 Systematic Dissection of the Areas Proximal to the Rearranged V_κ Gene

Our laboratory is interested in determining whether there are sequences in, or proximal to, the rearranged gene that are required for hypermutation. We approached the problem by systematically deleting the 5′ non-coding region, the VJ coding region, and the 3′ non-coding region of the V_κ167 gene and by moving the promoter upstream of its original site by two kilobases (Fig. 1). In the first construct, the 5′ region (440 bp) between the promoter and the first codon was removed. This included the leader sequence and the leader intron, but not the 5′ cap site where transcription starts. In the second construct, we removed a fragment which started at the first codon and ended at the end of J_κ5, thereby deleting the entire coding region. In the third construct, we removed a 750 bp fragment which started just downstream of the J_κ5 and progressed to the intronic *SacI* site. In the

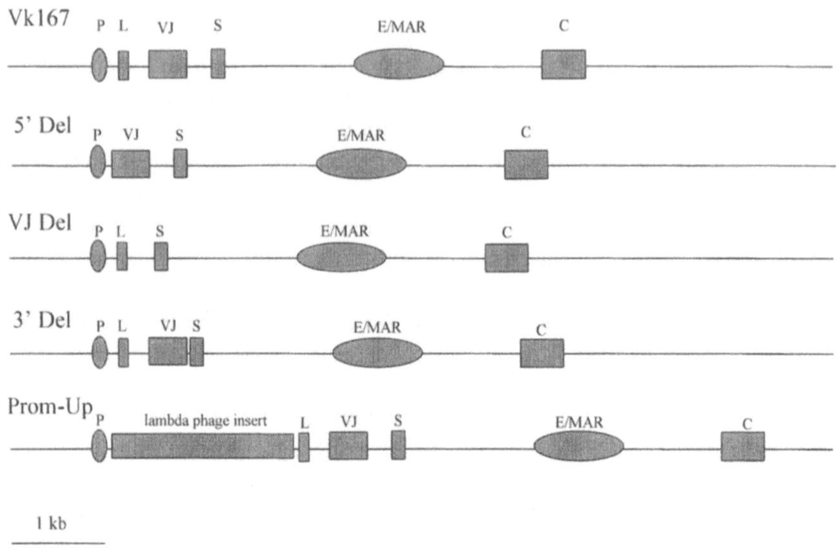

Fig. 1. Deletion and insertion constructs derived from the V_κ167 light chain. The V_κ167 light chain, utilized in the anti-phosphorycholine response, was used to make a transgene as previously described (Umar et al. 1991). This transgene was modified by removing or inserting segments of DNA as illustrated above. In the *5′ Del* construct, 440 bases were removed starting from just downstream of the transcription initiation site and ending at codon 1 of the variable gene. In the *VJ Del* construct, the variable coding region (330 bp) was removed starting at codon 1 and ending at the last codon of the J region. In the *3′ Del* construct, 750 bases were removed, starting just after the J region and ending at the *SacI* site in the J-C intron. The *Prom-Up* construct was produced by inserting 2 kb of λ phage DNA just downstream of the transcription initiation site. All of the constructs included the *supF* tRNA gene added to the J-C intron and the 3′ κ enhancer required for hypermutation (not shown in the figure). *P*, promoter; *L*, leader sequence; *VJ*, rearranged variable coding region exon; *S*, *supF* tRNA gene; *E/MAR*, J-C intronic enhancer and matrix attachment region; *C*, constant region exon

final construct, we inserted 2 kb of λ phage DNA between the promoter and the leader sequence, just downstream of the 5′ cap site. This would allow us to determine if proximity of the promoter to the VJ region was important for mutation. All of the constructs also included the *supF* tRNA gene inserted in the J-C intron to act as a screening marker and 3′ transgene-specific PCR anchor. Also, because of the placement of the deletions or insertions, all of the transgenes retained the capacity to transcribe, but produced nonfunctional transcripts which would not interfere in selection of B cells for a protein product.

Measuring the frequency of mutation in the constructs was accomplished by screening immunized transgenic mice. The transgenes were amplified by PCR and cloned from DNA isolated from $B220^+$, PNA^+ B cells from the transgenic mice. The clones were then screened for mutation by the single-stranded confirmation polymorphism (SSCP) assay. While the data reported here on the deletion constructs is preliminary, it already shows a strong trend. Both the 5′ deletion and VJ deletion constructs are still able to undergo mutation, but the 3′ deletion construct undergoes little or no mutation (Table 2). These data agree with earlier reports

Table 2. Frequency of mutated clones found by SSCP in deletion constructs

Deletion construct	Mutant clones (%)
5′ Del	4.6 (13/303)
VJ Del	3.1 (5/160)
3′ Del	< 0.6 (0/160)

(Betz et al. 1994; Yélamos et al. 1995) that neither the 5′ region nor the VJ coding region are necessary for targeting the hypermutation mechanism, but suggests that the area immediately 3′ of $J_\kappa 5$ does play an essential role.

The construct in which the promoter was moved 2 kb upstream of its original position was tested in a similar manner to the other transgenes. It was inserted into the genome of three founder lines of mice, checked to ensure that it underwent transcription, and tested for mutation after antigenic challenge (Winter et al. 1997). We screened both the normal epicenter of mutation (VJ) and the region of the phage DNA insert just downstream of the promoter (see Fig. 1). Because the fragment of interest was much larger than in the other constructs, we used Taq polymerase rather than Pfu polymerase to amplify it from the $B220^+$, PNA^+ B cells. This resulted in a much higher background of frequency of mutations due to polymerase error, which was corrected for by subtracting out the background frequency of mutation from the same regions of the transgene amplified from tail DNA, which presumably does not hypermutate. After screening 3×10^5 bases (approximately 300 clones) for mutation, we found no hypermutation in either the VJ region or in the λ phage region (Table 3). Endogenous, rearranged V_H genes from the same pool of B cells, however, underwent normal levels of hypermutation. There are two possible conclusions drawn from the data: either the λ phage DNA insert is not a viable substrate for hypermutation, or the proximity of the promoter to other elements in or near the VJ coding region is important.

It is possible that the phage DNA insert is not recognizable as a substrate for mutation because it is too different from the variable sequence or has too many CpG methylation sites. This is unlikely since our data, reported above, and data from other laboratories (Yélamos et al. 1995; Peters and Storb 1996) have demonstrated that the VJ coding region is not required for mutation. Furthermore, it has been demonstrated that prokaryotic DNA fragments placed downstream of the promoter can act as substrates for mutation (Yélamos et al. 1995). However, these

Table 3. Frequency of mutated clones found by SSCP in prom-up transgenic mice

Source of DNA	Mutated clones (%)
λ phage transgene	0 (0/266)
$V_\kappa 167$ transgene	0 (0/253)
Endogenous $V_H S107$	24 (28/101)

fragments were much shorter than our insert. This, in turn, supports the second conclusion – that close proximity of the promoter to elements in or around the VJ region is required for hypermutation.

Recent work from Tim Manser's laboratory has shown that the 5′ boundary of hypermutation moves upstream with the V_H promoter when 750 bp of *D. melanogaster* intronic DNA is inserted into the leader intron of a V_H transgene construct (Tumas-Brundage and Manser 1997). This work supports an earlier hypothesis that hypermutation tracks with the promoter rather than the variable gene (Lebecque and Gearhart 1990; Weber et al. 1994). While this data is at variance with ours, it may be explained by the difference in distance (750 bp vs. 2000 bp) and differences in requirements for mutation in V_H and V_L chains. Moving the promoter upstream by a shorter amount may not affect its interactions with other DNA elements around the VJ coding region, whereas moving it 2 kb may place it out of reach of downstream elements.

The results of our deletion experiments reported above, as well as the results of others (Betz et al. 1994; Yélamos et al. 1995), rules out the importance of promoter or 5′ region interacting with the VJ exon. The lack of hypermutation in our constructs – one missing the 3′ flanking sequence and the other moving the promoter 2 kb upstream – supports the notion that the promoter may be interacting with elements just downstream of the J region. Our laboratory is currently exploring this possibility.

4 Models of Hypermutation

There are currently three models in vogue to explain hypermutation in humans and mice: the replication-dependent error-prone repair model (Manser 1990; Rogerson et al. 1991), the DNA-RNA-DNA copying loop model (Steele and Pollard 1987), and the transcription-coupled error-prone repair model (Brenner and Milstein 1966; Lebecque and Gearhart 1990; Peters and Storb 1996). All three models evolved around the observation that hypermutation demonstrates strand bias; that is to say, the mutational machinery appears to be targeting one strand, and not both, for mutation (Golding et al. 1987). Furthermore, the models all require an error-prone polymerase. While there is no evidence that a polymerase is actually involved in the hypermutation mechanism, it is the most straightforward possibility.

In the replication-dependent model, the mutations are hypothesized to occur on one strand or the other during replication. The problems with this model include evidence that immunoglobulin transgenes may be inserted in either orientation in regards to an endogenous origin of replication and still undergo hypermutation (Rogerson et al. 1991). Furthermore, this model does not explain why hypermutation tracks with the promoter (Tumas-Brundage and Manser 1997)

In the DNA-RNA-DNA copying loop model, RNA polymerase and a hypothetical error-prone reverse transcriptase create mutated cDNA transcripts which are then integrated back into the variable gene locus, either directly or through gene conversion. This model predicts that the transcriptional unit would be the primary target of the mutator mechanism, which would explain the targeted area of mutation and the observed strand bias. The weaknesses of the model are the lack of a known error-prone reverse transcriptase in B cells and the fact that as many as 5% of the mutations are upstream of the transcriptional unit.

In the transcription-coupled repair model, gratuitous error-prone repair could occur during transcription. It might take advantage of the increased number of palindromes and inverted repeats in the region which could give rise to stem-loop structures (Golding et al. 1987). A strand-specific repair mechanism could mistake mismatches in the palindromes for errors and begin error-prone repair of the region. This model would also account for the requirement for transcriptional elements, but is weakened by the ability of substituted non-immunoglobulin sequences to undergo hypermutation and by mutations found upstream of the transcription initiation site. In addition, all three models are severely weakened by recent data demonstrating that hypermutation occurs in repair deficient mice and human repair-deficient patients (Wagner et al. 1996; Winter and Gearhart, unpublished data). This suggests that neither nucleotide excision repair nor mismatch repair appear to be directly involved in inserting mutations during error-prone repair. Furthermore, in the mismatch repair deficient mice, there is less strand bias. (Winter and Gearhart, unpublished data). This is evidence that the mutator mechanism may target both strands, but strand-specific repair normally repairs one strand before the mutations can be fixed into the genome by replication. If this is true, then all the current models must be rethought.

5 Stalled Transcription–Replication Model for Hypermutation

If we take into account the most recent data reported here and by others, then a model for the hypermutation mechanism must fulfill the following requirements: specificity for the variable region of rearranged immunoglobulin genes; necessity of the E/MAR and the 3′ E; the ability of the epicenter of mutation to move upstream with the promoter; the ability of other DNA sequences to undergo mutation if substituted for the variable coding region; the necessity for close proximity of the promoter to elements just downstream of the V(D)J coding region; the ability to function without the help of error-prone mismatch repair or error-prone nucleotide excision repair; and the ability to mutate both strands. None of the current models of hypermutation can resolve all of the obstacles placed by these requirements, so a new hypothesis must be shaped around the available data.

We would propose a model with the following attributes. The E/MAR and the 3′ E target the immunoglobulin locus to a specific region of the nuclear matrix which houses both transcriptional machinery and mutational machinery (Fig. 2). Here the immunoglobulin genes can undergo regulated transcription. Once the hypermutational machinery is turned on, a mutational factor (XF) interacts with high affinity for both the transcription initiation elements (i.e. TFIID, TFIIB, or RNA polymerase) in the promoter region and the area just 3′ of the J gene segments. This binding serves two functions: It halts RNA transcription in the initiation state, therefore physically blocking the area 5′ of the immunoglobulin gene against hypermutation; and it destabilizes the DNA within the loop. If replication begins at this time from a cryptic origin of replication near or in the E/MAR

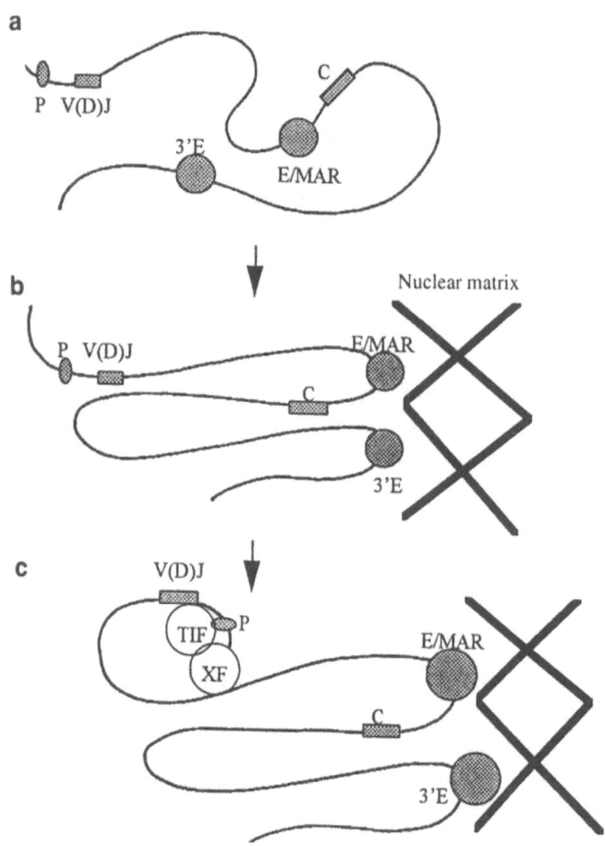

Fig. 2a–c. Stalled transcription–replication model of hypermutation. **a** The rearranged immunoglobulin locus remains quiescent until, in **b**, its enhancers associate with an Ig specific site on the nuclear matrix which up-regulates its transcription. **c** Upon B cell entrance into the germinal center and the proper external signal, an unknown mutation factor (*XF*), associates with the transcription initiation factors (*TIF*), stalling them in place at the 5′ end of the gene. It also associates either directly or through other trans-acting factors to sequences in the J-C intron, most probably just downstream of the J region. This loops out the region in between the promoter and the enhancer/matrix attachment region (*E/MAR*). Replication initiated at this time from a cryptic origin of replication in the J-C intron would be stalled at this loop and errors are introduced into the leading and lagging strands by DNA polymerase

region, then replication down both strands towards the immunoglobulin promoter is inhibited. The closer the replication fork comes to the promoter, the more inhibited it becomes, and the stalled DNA polymerase inserts the wrong nucleotides on the leading and lagging strand. The majority of the time this activity would halt when it runs into the physical barrier imposed by the stalled transcription initiation complex, about 50 bp downstream of the transcription start site. On occasion it might force its way past, allowing for the infrequent mutations found upstream of the transcriptional unit.

This model would satisfy the requirements for gene specificity, the role of enhancers and promoter, the interaction between promoter and downstream elements, the ability of substituted DNA sequences to undergo mutation in place of the V(D)J, and the lack of strand bias. Furthermore, centroblasts may be especially prone to replication-linked repair since they are rapidly dividing (approximately every 8 h) in germinal centers. Like the other models, this one has its weaknesses, in that it requires a hypothetical cryptic origin of replication within the J-C intron. Eucaryotes have a much more complex system of origins of replication than do procaryotes, with an origin placed about every 800 bases, often near matrix attachment regions. It is therefore quite possible that there is a previously undescribed origin of replication like the one we have proposed and we are currently testing the region for such origins.

We are not completely satisfied with our own model or the models of others. Too many pieces of the puzzle are missing; other pieces are contradictory, as if they came from another puzzle box. But as inadequate as any of our models are, they give us a place to stand, a way to compare data with hypothesis and a starting point from which to build better models.

References

Betz AG, Milstein C, Gonzalez-Fernandez A, Pannell R, Larson T, Neuberger MS (1994) Elements regulating somatic hypermutation of an immunoglobulin κ gene: critical role for the intron enhancer/matrix attachment region. Cell 77:1–10

Brenner S, Milstein C (1966) Origin of antibody variation. Nature 211:242–243

Close PM, Pringle JH, Ruprai AK, West KP, Lauder I (1990) Zonal distribution of immunoglobulin-synthesizing cells within the germinal centre: an in situ hybridization and immunohistochemical study. J Pathol 162:209–216

Dianda L, Gulbranson-Judge A, Pao W, Hayday AC, MacLennan ICM, Owen MJ (1996) Germinal center formation in mice lacking αβ T cells. Eur J Immunol 26:1603–1607

Gearhart PJ (1993) Somatic mutation and affinity maturation. In: Paul WE (ed) Fundamental immunology. Raven, New York, pp865–885

Gearhart PJ, Bogenhagen DF (1983) Clusters of point mutations are found exclusively around rearranged antibody variable genes. Proc Natl Acad Sci USA 80:3439–3443

Gearhart PJ, Johnson ND, Douglas R, Hood L (1981) IgG antibodies to phosphorylcholine exhibit more diversity than their IgM counterparts. Nature 291:29–34

Giusti AM, Manser T (1993) Hypermutation is observed only in antibody H chain V region transgenes that have recombined with endogenous immunoglobulin H DNA: implications for the location of cis-DNA elements required for somatic mutation. J Exp Med 177:797–809

Golding GB, Gearhart PJ, Glickman BW (1987) Patterns of somatic mutations in immunoglobulin variable genes. Genetics 115:169–176
Green NS, Sack SZ (1997) Effect of the Bloom's syndrome DNA helicase on Ig hypermutation. Keystone Symposium on B Lymphocytes in Health and Disease. Abstract 420
Hackett J, Rogerson BJ, O'Brien RL, Storb U (1990) Analysis of mutations in κ transgenes. J Exp Med 172:131–137
Lebecque SG, Gearhart PJ (1990) Boundries of somatic mutation in rearranged immunoglobulin genes: 5′ boundary is near the promoter, and 3′ boundary is ~1 kb from V(D)J gene. J Exp Med 172:1717–1727
Liu Y-J, Johnson GD, Gordon J, MacLennan ICM (1992) Sites of specific B cell activation in primary and secondary responses to T cell-dependent and T cell-independent antigens. Immunol Today 13:17–21
Manser T (1990) The efficiency of antibody affinity maturation: can the rate of B-cell division be limiting? Immunol Today 11:305–308
Matsumoto M, Lo SF, Carruthers CJL, Min J, Mariathasan S, Haung G, Plas DR, Martin SM, Geha RS, Nahm MH, Chaplin DD (1996) Affinity maturation in germinal centres in lymphotoxin-α-deficient mice. Nature 382:462–466
Meyer KB, Neuberger MS (1989) The immunoglobulin κ locus contains a second, stronger B-cell-specific enhancer which is located downstream of the constant region. EMBO J 8:1959–1964
Peters A, Storb U (1996) Somatic hypermutation of immunoglobulin genes is linked to transcription initiation. Immunity 4:57–65
Reynaud CA, Garcia C, Hein WR, Weill J-C (1995) Hypermutation generating the sheep immunoglobulin repertoire is an antigen-independent process. Cell 80:115–125
Rogerson B, Hackett J Jr, Peters A, Haasch D, Storb U (1991) Mutation pattern of immunoglobulin transgenes is compatible with a model of somatic hypermutation in which targeting of the mutator is linked to the direction of DNA replication. EMBO 10:4331–4341
Selsing E, Storb U (1981) Somatic mutation of immunoglobulin light-chain variable-region genes. Cell 25:47–58
Sharpe M, Milstein C, Jarvis JM, Neuberger MS (1991) Somatic hypermutation of immunoglobulin κ may depend on sequences 3′ of C_κ and occurs on passenger transgenes. EMBO J 10:2139–2145
Sohn J, Gerstein RM, Hsieh C-L, Lemer M, Selsing E (1993) Somatic hypermutation of an immunoglobulin μ heavy chain transgene. J Exp Med 177:493–504
Steele EJ, Pollard JW (1987) Hypothesis: somatic hypermutation by gene conversion via the error prone DNA → RNA → DNA information loop. Mol Immunol 24:667–673
Texido G, Jacobs H, Meiering M, Kuhn R, Roes J, Muller W, Gilfillan, Fugiwara H, Kikutani H, Yoshida N, Amakura R, Benoist C, Mathis D, Kishimoto T, Mak TW, Rajewsky K (1996) Somatic hypermutation occurs in B cells of terminal deoxynucleotidyl transferase-, CD23-, interleukin-4-, IgD-, and CD30-deficient mouse mutants. Eur J Immunol 26:1966–1969
Tumas-Brundage K, Manser T (1997) Transcriptional promoter regulates hypermutation of the antibody heavy chain locus. J Exp Med 185:239–250
Umar A, Schweitzer PA, Levy NS, Gearhart JD, Gearhart PJ (1991) Mutation in a reporter gene depends on proximity to and transcription of immunoglobulin variable transgenes. Proc Natl Acad Sci USA 88:4902–4906
Wagner SD, Elvin JG, Norris P, McGregor JM, Neuberger MS (1996) Somatic hypermutation of Ig genes in patients with xeroderma pigmentosum (XP-D). Int Immunol 8:701–705
Weber JS, Berry J, Manser T, Claflin JL (1994) Mutations in Ig V(D)J genes are distributed asymmetrically and independently of the position of V(D)J. J Immunol 153:3594–3602
Winter DB, Sattar N, Mai J-J, Gearhart PJ (1997) Insertion of 2 kb of bacteriophage DNA between an immunoglobulin promoter and variable region stops somatic hypermutation in a κ transgene. Molec Immunol 34:359–366
Wykes M, Poudrier J, Lindstedt R, Gray D (1997) Soluble CD40 ligand production by mouse B cells. Keystone Symposium on B Lymphocytes in Health and Disease. Abstract 253
Yélamos J, Klix N, Goyenechea B, Lozano F, Chui YL, Gonzalez-Fernandez A, Pannell R, Neuberger MS, Milstein C (1995) Targeting of non-Ig sequences in place of the V segment by hypermutation. Nature 376:225–229
Zheng B, Xue W, Kelsoe G (1994) Locus-specific somatic hypermutation in germinal centre T cells. Nature 372:556–559

Somatic Hypermutation of Immunoglobulin Genes is Linked to Transcription

U. Storb[1,3], A. Peters[2], E. Klotz[3], N. Kim[2], H.M. Shen[1], K. Kage[1], B. Rogerson[4] and T.E. Martin[1,3]

1 Introduction

Immunoglobulin (Ig) genes are rearranged in pre-B cells. Pre-B cells that express Ig heavy (H) and light (L) chain genes whose V(D)J recombination results in a functional reading frame mature into B cells that exit the bone marrow. The V(D)J recombination process creates a large repertoire of different variable regions from a restricted pool of germline genes. Additional variablity arises during the process of somatic hypermutation in mature B cells proliferating in germinal centers of lymphoid organs (reviewed in French et al. 1989). B cells that have mutated to express high-affinity antibodies are selected and develop into plasma cells or memory cells. B cells with mutations that decrease the affinity of the expressed Igs or that prevent Ig expression die by apoptosis. The somatic point mutations are located within the variable region and their proximate upstream and downstream flanks, but not generally within the constant region.

[1]Department of Molecular Genetics and Cell Biology, University of Chicago, Chicago, IL 60637, USA
[2]Department of Biochemistry and Molecular Biology, University of Chicago, Chicago, IL 60637, USA
[3]Committee on Immunology, University of Chicago, Chicago, IL 60637, USA
[4]Trudeau Institute, Saranac Lake, NY 12983, USA

The molecular basis of somatic mutation of Ig genes is not known. Many hypotheses have been formulated (discussed in STORB 1996). This chapter reviews the evidence for a new model that links somatic mutation to transcription.

2 Ig Transgenes are Useful Substrates for Analysis of the Cis-acting Sequences Involved in Somatic Hypermutation

Ig transgenes were first studied for somatic mutation to determine if V(D)J recombination plays a role in the mutation process (O'BRIEN et al. 1987). It was found that rearranged Ig transgenes are mutable, thus showing that V(D)J recombination is not required and establishing the transgenic system as an experimental tool for the study of somatic mutation. Mutations occur in Ig transgenes regardless of the site of integration in the genome, demonstrating that sequences that regulate the mutation process in cis are present within the relatively short transgenes (O'BRIEN et al. 1987). The distribution of the mutations is as in endogenous genes, only V regions and their immediate flanks are mutated, but not the C region (HACKETT et al. 1990). Also, as in endogenous Ig genes, the mutations are primarily point mutations, with very rare deletions, and show a strong preference for nucleotide transitions over transversions (HACKETT et al. 1990; ROGERSON et al. 1991a,b).

Experiments with Ig transgenic mice have since been carried out in several laboratories, supporting the original findings (SHARPE et al. 1991; GIUSTI and MANSER 1993; SOHN et al. 1993). It was found for κ genes that the 3′ enhancer is required for maximal frequency of mutations, but that the basic mechanism proceeds normally in the absence of this enhancer (BETZ et al. 1994). On the other hand, deletion of the κ-intron enhancer, despite an intact 3′ enhancer, did not permit detectable somatic mutation. Thus, the κ-intron enhancer is sufficient and seems to be required (see below) for somatic mutation of κ genes. The heavy chain intron enhancer, as well as the λ enhancer are sufficient for somatic mutation (SOHN et al. 1993; KLOTZ and STORB 1996). Susprisingly, the Ig promoter is not required for high levels of somatic mutation as it can be replaced by a β-globin promoter (BETZ et al. 1994).

Transgenes most often are integrated in tandem, as multiple copies. Several or all of the multiple copies can be targeted by somatic mutation (SHARPE et al. 1991). However, in some cases only some of the copies are mutated (O'BRIEN et al. 1987; SHARPE et al. 1991). In one detailed study of a three-copy transgene array, one of the copies was targeted with a very high preference over the other two copies (ROGERSON et al. 1991a). Contrary to the published interpretation of this finding (ROGERSON et al. 1991a), it appears likely now that the preferential targeting was due to preferential transcription of the highly targeted transgene copy (STORB et al. 1996).

While naturally the major targets of the somatic mutation process are the V(D)J regions of Ig genes, there seems to be little, if any target specificity when unrelated sequences replace or interrupt the V region. A transgene consisting only of the heavy chain promoter and enhancer driving the expression of a bacterial chloramphenicol acetyl transferase (CAT) gene was mutated in the CAT region (AZUMA et al. 1993). Also, the bacterial gpt and neo genes, as well as a β-globin sequence, a completely artificial test substrate, and the C region were targeted by the somatic mutation process when located in the context of a κ gene within less than 1.5 kb of the start of transcription (YÉLAMOS et al. 1995; PETERS and STORB 1996; STORB et al. 1996).

A useful technical advance for the analysis of somatic mutation in both transgenes and endogenous genes has been the isolation of Peyer's patch or spleen B cells expressing high levels of peanut agglutinin (PNA) (GONZALES-FERNANDEZ and MILSTEIN 1993; ROGERSON 1995). These cells represent a mixture of B cells that are in the process of mutating and recently mutated cells.

3 Transcript Initiation Within an Ig Gene Causes Downstream Mutations

In order to determine if transcription played a role in somatic mutation we created a κ transgene whose transcriptional promoter was duplicated [together with the leader (L) sequence] and placed upstream of the C region (PETERS and STORB 1996). B cells of these mice showed similar levels of two different mature transcripts from the transgene. One transcript initiated as usual from the promoter upstream of the V region; it contained the spliced L-VJ-C sequences and terminated in poly(A). The other one initiated at the internal promoter upstream of the C region and thus contained L-C and terminated in poly(A). Somatic mutations were found in both the VJ region and the C region, but not in the intron between these. The data strongly suggest that somatic mutation is linked to transcription initiation and concentrated in a limited region downstream of the promoter.

4 Somatic Mutation Shows a DNA Strand Bias and an A/T Bias

Considering the nucleotide changes due to somatic mutation, evidence for a strand bias has been noted (GOLDING et al. 1987; NEUBERGER and MILSTEIN 1995; SMITH et al. 1996). An example is shown in Table 1. To avoid a selection bias, the sequences are from non-coding regions. To avoid a bias based on the base composition of the sequences, the data are corrected for the frequency at which each nucleotide occurs in the sequence. As is conventionally done, the data are represented as changes in the non-transcribed, i.e., coding strand (the top strand in conventional writing of double stranded DNA). As can be seen, changes from A are

Table 1. Normalized frequencies for each type of single-base change in somatically mutated Ig genes[a] [from SMITH et al. (1996)]

Substitution	A (%)	B (%)	C (%)	D (%)
G > N	25.02	23.44	24.07	21.03
C > N	22.26	27.98	25.97	24.65
A > N	34.63	33.28	33.27	38.67
T > N	**15.78**	**15.27**	**15.59**	**14.75**
Mutation number	341	323	664	179
Frequency	2.01%	0.51%	0.82%	1.55%

[a]Mutations for each mutating nucleotide (not including deletions).
A, C, D, SMITH et al. (1996); B, CLARKE et al. (1982); LEBEQUE and GEARHART (1990); ROTHENFLUTH et al. (1993); WEBER et al. (1994).

at least twice as frequent as changes from T in this sample. This bias has been observed in the other studies (GOLDING et al. 1987; NEUBERGER and MILSTEIN 1995). There is often also a differential for changes from G vs C; however, the difference is generally smaller and can go either way. The A > T difference was not seen for meiotic mutations of pseudogenes (GOLDING et al. 1987). The A > T preference has been interpreted as a strand bias, indicating that either of the two DNA strands is preferentially, or exclusively mutated. However, as shown in Table 2, a strand bias alone would not be detectable by inspecting the nucleotide changes, unless there is also an A/T bias.

If A and T are equally mutable, then, even if there is a strand bias, the frequency of mutations in A should be equal to the mutations in T when the data are corrected for base composition. Thus, a strand bias could not be detected by inspecting the mutability of A and T. Likewise, if the mutation rate is unequal for A and T, but no strand bias exists, the mutation frequency for A would be the same as for T, because for any A or T mutated in either strand, the top strand would show a change.

Table 2. Somatic mutation strand bias and A/T bias

A,T Changes expected in top strand[a]
If there is only a strand bias expect A = T
If there is only an A/T bias expect A = T
If there is both a strand bias and an A/T bias:
Top strand bias
Mutation rate A > T expect A > T[b]
Mutation rate T > A expect T > A
Bottom strand bias
Mutation rate A > T expect T > A
Mutation rate T > A expect A > T[b]

[a]A = T, the same proportion of A and T nucleotides are mutated. A > T (T > A), a higher proportion of A than T (T than A) nucleotides are mutated.
[b]The observed mutations are A > T when inspecting the top strand. This can only be explained when one assumes both an A/T bias and a strand bias. Thus, only the situations marked[b] lit the data.

If, however, there is both a strand bias and an A/T bias, changes will be seen depending on the strand and the nucleotide (A or T) that is preferentially mutated. Since, when inspecting the top strand sequence, A is mutated more frequently than T, only the two mutually exclusive situations, mutation rate A > T in the top strand bias and mutation rate T > A in the bottom strand bias, marked with *b* in Table 2, are possible. If the top strand is the target for mutations, A must be preferentially mutated; however, if the bottom strand is the target, T must be preferred (this would of course appear as an A > T bias when the top strand sequence is displayed). Thus, if we assume that the link to transcription indicates that the transcribed (bottom) strand is mutated, T would be more likely to be a target than A. This combination of a strand bias and A/T bias may eventually be exploited to understand the mechanism underlying somatic mutation. One possibility is that the bias is related to the strength of pairing of the specific ribonucleotides (rN) and deoxyribonucleotides (dN) involved. The r/d pair rU/dA has a stability about three times lower than rA/dT (Martin and Tinoco 1980). Thus, rU/dA < rA/dT; rG/dC =~ rC/dG. If mutations arise on the DNA strand that is being transcribed, a certain stability of r/d pairing may be required to introduce the mutations. Ig gene somatic mutations show a dearth of dA changes (on the transcribed DNA strand; see Table 1). This could support a model where the newly transcribed RNA, when strongly paired with the DNA, favors action of the mutator factor, leading to somatic mutation.

In any case, the data support a strand bias. This would be expected if the mutation process is directly linked to transcription, since only one of the two strands is transcribed. A strand bias is also compatible with DNA replication if the mutation process were biased for either the leading or the lagging strand coming from a unique origin of replication (Rogerson et al. 1991; Storb 1996). However, the data linking somatic mutation to transcription (Peters et al. 1996) make a replication model less likely. Finally, several mechanisms of DNA repair have been shown to preferentially repair the transcribed strand (Friedberg et al. 1995).

5 Transcription-Linked Model of Somatic Mutation

Since initiation of Ig gene transcription appears to result in somatic mutation in mutating B cells, it is likely that a mutator factor is translocated into the Ig gene by the act of transcription (Peters and Storb 1996). It appears that such a factor must bind to the initiating RNA polymerase and act during transcript elongation for up to 1.5 kb from the start (Storb 1996).

We have proposed a model of transcription-coupled DNA repair (Peters and Storb 1996). In this model, a mutator factor is bound to the RNA polymerase at the promoter and remains associated with the polymerase during elongation. The mutator factor would cause stalling of the polymerase or prevent resolution of naturally occuring polymerase pausing. This would call into action a DNA repair system which would cause the excision of a small region of the transcribed DNA strand. DNA polymerization of that region would result in occasional errors (even if a high-fidelity DNA polymerase, such as δ or ε, were involved) which would be retained in at least one of the daughter cells after the next replication step. The mutations would be seen only over the first 1.5 kb or so of the Ig gene, because the stalling event would occur no farther than 1.5 kb from the transcription start during each round of transcription with the factor loaded. Once stalling had occurred, the factor would fall off the transcription complex and be unable to reload on the elongating complex after resolution of the stop. The factor could of course load onto newly initiating complexes. Alternatively, the factor would have only a limited affinity for the polymerase and dissociate within 1.5 kb, whether a stalling-induced mutation had occurred or not.

One candidate for a repair system involved would be nucleotide excision repair (NER). This has now been tested by others (Wagner et al. 1996) and ourselves (Kim et al. 1997; Shen et al. 1997) with B cells from patients or mice defective in NER. It appears that B cells with a defect in one of these genes are still able to somatically mutate Ig genes.

Mismatch repair has recently been shown to also preferentially target the transcribed strand (Mellon et al. 1996). It is conceivable that this type of repair is induced during transcription of Ig genes or stalling of transcripts. Since the unmutated Ig gene does not have a mismatch, one could envision that the mutS homolog of mismatch repair would be replaced by a protein specific to somatic mutation which would bind to the DNA and induce the subsequent steps of mismatch repair.

The challenge is now to determine how transcription is linked to somatic mutation and whether indeed a known or novel DNA repair system is involved, and finally, to identify and characterize the mutator factor.

6 Is Somatic Mutation in B Cells Ig Gene Specific?

In all transgenic (and endogenous) Ig genes so far shown to undergo somatic hypermutation, the genes contained an Ig enhancer. However, the Ig promoter appears not to be required, as it can be replaced by a β-globin promoter (Betz et al. 1994). The question then arises, whether somatic mutation is truly Ig gene specific, or if other genes that are transcribed at the time when the somatic mutation process is active can likewise be targeted. Most Ig gene enhancers have been found sufficient to allow mutations, with the exception of the 3′ κ enhancer (see "Ig Transgenes are Useful Substrates for Analysis of the Cis-acting Sequences Involved in Somatic Hypermutation"). The 3′ κ enhancer, however, is rather homologous to the λ enhancers (Pongubala et al. 1992; Eisenbeis et al. 1995), that are sufficient for somatic mutation (Klotz and Storb 1996). The κ transgene that contained only the 3′ enhancer had a deletion of the κ intron enhancer, but the 3′ enhancer was left in its original position, 3′ of Cκ (BETZ et al. 1994). It is not clear if there may be a position or distance effect. Perhaps, if the 3′ enhancer were placed in the position of the κ intron enhancer somatic mutation would occur.

Assuming then that all Ig gene enhancers can support somatic mutation in combination with Ig promoters or non-Ig promoters, is there really Ig gene specificity of the process? The Ig enhancers are different from each other and share motifs with other enhancers. It is thus possible that all active genes are mutated in B cells undergoing somatic mutation. We have started to investigate this question by sequencing housekeeping genes in B cells which have highly mutated their Ig genes (PETERS and STORB, unpublished data). Our very preliminary results indicate that housekeeping genes are not mutated, the limit being at least 100-fold lower than the Ig genes. If confirmed, this would suggest that the Ig enhancer is responsible for the Ig gene specificity of somatic hyprmutation (Fig. 1).

7 Conclusions

The currently available data support the following model for somatic mutation of Ig genes. A mutator factor is active only in mutating B cells. Interaction of the transcriptional promoter with an Ig enhancer allows the binding of this factor specifically to RNA polymerases associated with Ig genes. The factor remains bound to the polymerase during transcript elongation up to about 1.5 kb from the promoter and causes point mutations by an unknown process, perhaps linked to DNA repair.

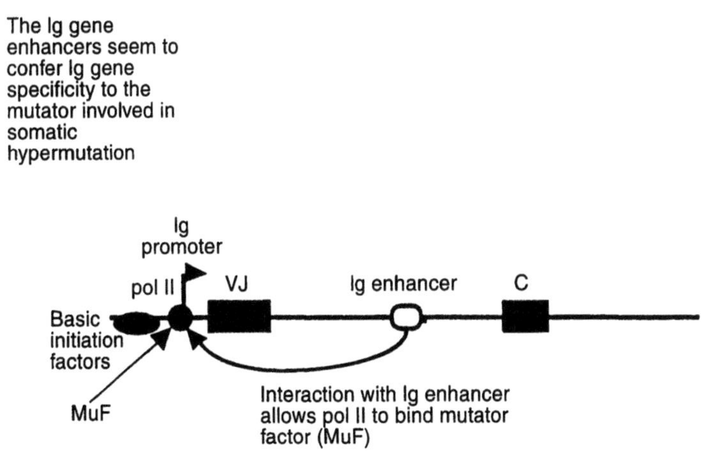

Fig. 1. The Ig gene enhancers seem to confer Ig gene specificity to the mutator involved in somatic hypermutation

Acknowledgements. The work from our laboratory has been supported by NIH grant GM38649.

References

Azuma T, Motoyama N, Fields L, Loh D (1993) Mutations of the chloramphenicol acetyl transferase transgene driven by the immunoglobulin promoter and intron enhancer. Int Immunol 5:121–130

Betz A, Milstein C, Gonzalez-Fernandez R, Pannell R, Larson T, Neuberger M (1994) Elements regulating somatic hypermutation of an immunoglobulin κ gene: critical role for the intron enhancer/matrix attachment region. Cell 77:239–248

Eisenbeis C, Singh H, Storb U (1995) Pip, a novel IRF family member, is a lymphoid-specific, PU.1-dependent transcriptional activator. Genes Dev 9:1377–1387

French D, Laskov R, Scharff M (1989) The role of somatic hypermutation in the generation of antibody diversity. Science 244:1152–1157

Friedberg E, Walker G, Siede W (1995) DNA repair and mutagenesis. American Society for Microbiology, Washington DC

Giusti A, Manser T (1993) Hypermutation is observed only in antibody H chain V region transgenes that have recombined with endogenous immunoglobulin H DNA: implications for the location of cis-acting elements required for somatic mutation. J Exp Med 177:797–809

Golding G, Gearhart P, Glockman B (1987) Patterns of somatic mutation in immunoglobulin variable genes. Genetics 115:169–176

Gonzales-Fernandez A, Milstein C (1993) Analysis of somatic hypermutation in mouse Peyer's patches using immunoglobulin κ light-chain transgenes. Proc Natl Acad Sci USA 90:9862–9866

Hackett J, Rogerson B, O'Brien R, Storb U (1990) Analysis of somatic mutations in κ transgenes. J Exp Med 172:131–137

Kim N, Kage K, Matsuda F, Lefranc MP, Storb U (1997) B lymphocytes of xeroderma pigmentosum or Cockayne syndrome patients with inherited defects in nucleotide excision repair are fully capable of somatic hypermutation of immunoglobulin genes. J Exp Med 186:413–419

Klotz E, Storb U (1996) Somatic hypermutation of a λ-2 transgene under the control of the λ enhancer or the heavy chain intron enhancer. J Immunol 157:4458–4463

Martin F, Tinoco I (1980) DNA–RNA hybrid duplexes containing oligo(dA:rU) sequences are exceptionally unstable and may facilitate termination of transcription. Nucleic Acids Res 8:2295–2299

Mellon I, Rajpal D, Koi M, Boland C, Champe G (1996) Transcription-coupled repair deficiency and mutations in human mismatch repair genes. Science 272:557–560

Neuberger MS, Milstein C (1995) Somatic hypermutation. Curr Opin Immunol 7:248–254

O'Brien R, Brinster R, Storb U (1987) Somatic hypermutation of an immunoglobulin transgene in κ transgenic mice. Nature 326:405–409

Peters A, Storb U (1996) Somatic hypermutation of immunoglobulin genes is linked to transcription initiation. Immunity 4:57–65

Pongubala J, Nagulapalli, Klemsz S, McKercher S, Mak R, Atchison M (1992) PU.1 recruits a second nuclear factor to a site important for immunoglobulin κ 3′ enhancer activation. Mol Cell Biol 12:368–378

Rogerson B (1995) Somatic hypermutation of VHS107 genes is not associated with gene conversion among family members. Int Immunol 7:1225–1235

Rogerson B, Hackett J, Peters A, Haasch D, Storb (1991a) Mutation pattern of immunoglobulin transgenes is compatible with a model of somatic hypermutation in which targeting of the mutator is linked to the direction of DNA replication. EMBO J 10:4331–4341

Rogerson B, Hackett J, Storb U (1991b) Somatic hypermutation in transgenic mice. In: Steele EJ (ed) Somatic hypermutation in V-regions. CRC Press, Boca Raton, Florida, pp 115–127

Sharpe M, Milstein C, Jarvis J, Neuberger M (1991) Somatic hypermutation of immunoglobulin κ may depend on sequences 3′ of Cκ and occurs on passenger transgenes. EMBO J 10:2139–2145

Shen HM, Cheo D, Friedberg E, Storb U (in press) Inactivation of the XP-C gene does not affect somatic hypermutation or class switch recombination of immunoglobulin genes. Mol Immunol

Smith D, Creadon G, Jena P, Portanova J, Kotzin B, Wysocki L (1996) Di- and trinucleotide target preferences of somatic mutagenesis in normal and autoreactive B cells. J Immunol 156:2642–2652

Sohn J, Gerstein R, Hsieh C, Lemer M, Selsing E (1993) Somatic hypermutation of an immunoglobulin μ heavy chain transgene. J Exp Med 177:493–504
Storb U (1996) Molecular mechanism of somatic hypermutation of immunoglobulin genes. Curr Opin Immunol 8:206–214
Storb U, Peters A, Klotz E, Rogerson B, Hackett J (1996) The mechanism of somatic hypermutation studied with transgenic and transfected target genes. Semin Immunol 8:131–140
Wagner S, Elvin J, Norris P, McGregor J, Neuberger M (1996) Somatic hypermutation of Ig genes in patients with xeroderma pigmentosum (XP-D). Int Immunol 8:701–705
Yélamos J, Klix N, Goyenechea B, Lozano F, Chui YL, Gonzalez-Fernandez A, Pannell R, Neuberger MS, Milstein C (1995) Targeting of non-Ig sequences in place of the V segment by somatic hypermutation. Nature 376:225–229

On the Possible Role of Natural Reverse Genetics in the V Gene Loci

R.V. BLANDEN[1], H.S. ROTHENFLUTH[1,2], and E.J. STEELE[1,2]

1 Introduction

The immune system of higher vertebrates has evolved to mount destructive responses which eliminate infectious agents. These responses depend upon a large population of mobile cells (lymphocytes) each of which expresses multiple copies of a particular receptor for antigen. The receptors are encoded by large multigene families in germline DNA, but before expression of receptor proteins in lymphocytes (each with a heterodimeric binding site for antigen) the germline genes undergo a unique rearrangement process in which two or three separate genetic elements are brought together to form the final coding sequence for the variable (V) portion of each of the two receptor protein chains (TONEGAWA 1983). Another separate element encodes the constant (C) region of the receptor protein which, in the case of the heavy chains of immunoglobulins and both chains of T cell receptors, spans the cell membrane and is an integral part of the signalling mechanism which activates lymphocyte responses to antigen. In the case of B lymphocytes, soluble immunoglobulins are secreted which have the same antigen-binding specificity as the receptor on each individual cell and which mediate effector functions through the constant region of the secreted antibody molecule.

The large number of germline V genes, the stochastic rearrangement of multiple elements to form the expressed V region, the heterodimeric nature of the binding site

[1]Division of Immunology and Cell Biology, John Curtin School of Medical Research, The Australian National University, PO Box 334, Canberra, ACT 2601, Australia
[2]Department of Biological Sciences, University of Wollongong, Northfields Avenue, Wollongong, New South Wales 2522, Australia

for antigen and additional capacity of the rearranged V gene in lymphocytes to undergo a high rate of mutation (STEELE 1991), all contribute to the immense potential diversity of the repertoire of receptors expressed on lymphocyte populations. This diversity enables the immune system to recognise virtually all foreign agents.

The system can only function, however, by virtue of deletion or suppression of lymphocytes with "faulty" or "dangerous" receptors which pose a threat to self by binding to self antigens, an inevitable consequence of the random processes which generate the repertoire and positive selection and proliferation of lymphocytes which recognise antigens. If other genes were subject to similar diversity-generating mechanisms to V genes, it would be biologically disastrous since the majority of amino acid sequence changes in most proteins adversely affect function.

This chapter concerns the mechanisms by which rearranged V genes undergo diversification by somatic hypermutation or gene conversion in B lymphocytes. We have approached this problem from an evolutionary perspective by posing the rhetorical question "What should evolution select to solve the adaptive immunity problem?" Our view of the critical issues with respect to somatic variation in antibody genes are listed in Table 1.

A review of available literature affords a description of what has been observed in the adaptive immune systems of higher vertebrates (STEELE et al. 1997). Firstly, there is the separation of the V and C coding elements (which allows diversity generation in V while preserving the essential signalling and effector functions of C). Secondly, separation of germline V coding elements into two or three sections and the assembly of the final V coding element by stochastic rearrangement of these elements, wholly or partly solves the problem of diversity of repertoire to enable identification of virtually all infectious agents.

2 Molecular Mechanism of Somatic Hypermutation of Rearranged V(D)J Genes in Lymphocytes

In our view, the only mechanism compatible with all available data involves reverse transcription to produce cDNA using pre-mRNA as a template, producing by these

Table. 1. Somatic hypermutation of antibody coding sequences. Theoretical Principles: What should evolution select?

1. Genetic separation of V and C coding regions so that V can mutate while C is conserved (to conserve effector function in H chain in particular)
2. A means of ensuring that mutation is confined to V coding regions so that effector functions of C regions are conserved and regulatory functions of promoters and enhancers are retained.
3. Mutation must not be permitted in other loci (this could have disastrous consequences for the cell and ultimately the animal)
4. Since most mutations lose affinity and only a small minority gain affinity, there would be selection for mechanisms:
   a) To select the high affinity mutants
   b) To stop them mutating further
   c) To expand their numbers by proliferation after they stop mutating.

error-prone copying mechanisms mutated cDNAs of V(D)J regions, which then undergo homologous recombination into chromosomal V(D)J DNA, thus replacing the original rearranged V gene (as first proposed by STEELE and POLLARD 1987). We postulate a "mutatorsome" (based on the precedent of telomerase, BLACKBURN 1992) which possesses reverse transcriptase activity and protein subunits which ensure that *only* rearranged V(D)J genes are mutated, thus avoiding the adverse consequences of mutation in any other genes. To achieve this locus specificity, we propose the binding of mutatorsome proteins to unique secondary structures in the pre-mRNA of immunoglobulin genes at a site in the intronic enhancer/ matrix attachment (Ei/MAR) region of the J-C intron of immunoglobulin heavy (H) and κ loci. Chromosomal location of the pre-mRNA mutatorsome complex could be achieved by binding to unique structures associated with the MAR region (e.g. DNA-RNA binding proteins). Downregulation of the genes encoding mutatorsome subunits would downregulate mutation.

We will first concentrate on available evidence concerning the mechanism of somatic hypermutation in the B lymphocytes of mice, much of which has been achieved with recombinant DNA technology to produce transgenic mouse lines with modified Ig loci. These data are summarised in Table 2 (discussed and analysed in greater detail in STEELE et al. 1997). The first ten points in Table 2 relate to the molecular mechanisms of mutation and points 11 and 12 are relevant to the cell biology of the germinal centres in which mutation takes place (BEREK et al. 1991; JACOB et al. 1991). Point mutation, the bias for transitions over transversions and strand bias are presumably signatures of the molecular mechanism of mutation. Stand bias is important because it indicates that the mechanism operates on only one DNA strand. This observation is compatible with reverse transcriptase-based models but seems difficult to reconcile with DNA repair-based models, given that known DNA repair mechanisms operate on double-stranded DNA and should not exhibit strand bias (although a model that could do so has been proposed by MANSER 1990).

The upstream and downstream boundaries of mutation, with the peak roughly focused on the rearranged V(D)J, is compatible with the RNA-based model outlined above but it is difficult to reconcile with DNA-based models, particularly with respect to points 6 and 7 in Table 2. The shape of the distribution of mutation described in point 5 of Table 2 is also compatible with the reverse transcriptase model. In this model, the distribution is accounted for by homologous recombination into the chromosomal rearranged V(D)J of varying lengths of mutated cDNA in different B lymphocytes (STEELE et al. 1992). The cDNAs with different lengths are provided by different starting points for cDNA synthesis on the pre-mRNA template according to primer length or nick site in alternative models described below and elsewhere (STEELE et al. 1997). This accounts for the gradually declining mutation frequency from the peak to the 3′ boundary within the J-C intron. The steep slope upstream of the peak is accounted for by a race between cDNA synthesis reaching the upstream end of the pre-mRNA template prior to ordered cutting of the template by splice mechanisms. If no splice cut occurs before cDNA progression through the L-V intron, the upstream boundary will be at or

Table 2. Features of somatic hypermutation of rearranged V(D)J genes in B lymphocytes of mice[a]

1.	Mainly point mutation, not gene conversion (CHIEN et al. 1988; GONZALEZ-FERNANDEZ and MILSTEIN 1993)
2.	Bias for transitions (59%) over transversions (BETZ et al. 1993)
3.	Strand bias i.e. the mutator mechanism operates on one DNA strand (BETZ et al. 1993)
4.	Mutation is roughly focussed on the V(D)J (LEBECQUE and GEARHART 1990; WEBER et al. 1991) a) Upstream boundary is transcription start site (V_H) or L-V intron (V_κ) (ROTHENFLUH et al. 1993; RADA et al. 1994; ROGERSON 1994) b) Downstream boundary is within J-C intron (V_H, V_κ) or downstream of C region (Vl) (BOTH et al. 1990; LEBECQUE and GEARHART 1990; MOTOYAMA et al. 1991)
5.	Distribution of mutations is asymmetric with a steep slope upstream of the peak and a more gradual slope downstream of the peak into the J-C intron (STEELE et al. 1992; ROTHENFLUH et al. 1993)
6.	The V(D)J coding sequence and all upstream sequence including the promoter can be exchanged for non-Ig sequence and mutation still occurs (AZUMA et al. 1993; YÉLAMOS et al. 1995; BETZ et al. 1994)
7.	The Ei/MAR region in the J-C intron is essential for mutation. Other downstream enhancers of transcription also enhance mutation (GIUSTI and MANSER 1993; SOHN et al. 1993; BETZ et al. 1994). A 300-base sequence spanning the A-T rich 3′MAR appears essential (analysis of data shown in HACKETT et al. 1992; AZUMA et al. 1993); the abundance of poly-A and poly-T sequences suggests a capacity for secondary RNA structure in the pre-mRNA
8.	When gene conversion is induced in transgene constructs, point mutation always occurs with it (XU and SELSING 1994)
9.	Transgenic constructs with a tRNA coding sequence inserted between the V(D)J and Ei/MAR regions generally do not allow mutation in the V(D)J (UMAR et al. 1991; UMAR and GEARHART 1995)
10.	Transgenic constructs with Ei/MAR downstream of the C region mutate in the C region (PETERS and STORB 1996)
11.	Most mutations lose affinity for epitope. A small minority (2–16%) gain affinity. (CHEN et al. 1992, 1995)
12.	Proliferation of B cells is regulated independently of mutation. Aged B cells do not mutate, but still proliferate in germinal centres (YANG et al. 1996)

[a] See STEELE et al. (1997) for more detailed analysis and referencing.

near the transcription start site, depending upon the activity of 5′ to 3′ exonuclease acting on the pre-mRNA template before the termination of cDNA synthesis. Alternatively, if the splice cut occurs at the upstream boundary of the L-V intron this will determine the upstream boundary of mutation, again dependent upon a race between splice cut and 5′ to 3′ exonuclease versus the completion of cDNA synthesis.

Points 6 and 7 in Table 2 are entirely compatible with the reverse transcriptase-based model presented here and previously, since this model invokes the Ei/MAR region in the J-C intron as an essential locus-specific device with mutation occurring upstream of this region, regardless of the nature of the upstream sequences, provided transcription occurs to produce the pre-mRNA template for reverse transcription. The concurrence of point mutation and gene conversion (point 8) is explained in our model by cDNA (containing point errors through reverse transcription) being the donor sequence in the gene conversion process.

Points 9 and 10 in Table 2 are compatible with most of the available evidence but require further definition by fully controlled experiments. At this time, most evidence available from the tRNA approach suggests that the insertion of a tRNA coding sequence between the V(D)J and Ei/MAR regions prevents the normal somatic hypermutation process from operating in the V(D)J. These data are compatible with a process of reverse transcription, operating in a 3′ to 5′ direction on the pre-mRNA template, being blocked by the secondary structure and modified bases of the tRNA insert, thus preventing production of the cDNA which replaces chromosomal V(D)J.

The excision of the Ei/MAR region and its relocation downstream of the C region, producing mutations in the C region, is also compatible with a critical focussing role for the Ei/MAR region and with a 3′ to 5′ direction of the mutation mechanism.

2.1 Priming of Reverse Transcription by Nicking of Chromosomal DNA

This priming mechanism for reverse transcription has been demonstrated in yeast (Zimmerly et al. 1995; Yang et al. 1996) and insects (Luan et al. 1993) and has been enunciated previously in detail in the context of somatic hypermutation (Steele et al. 1997). Briefly, the critical features of such a model is that nicks can be generated in either one or both strands of chromosomal DNA at specific sites distributed relatively uniformly along the J-C intron between the rearranged V(D)J and the Ei/MAR region (Fig. 1), thus generating a 3′ hydroxyl to act as a primer for reverse transcription of the pre-mRNA template which may anneal to the RNA-coding strand of the chromosomal DNA.

2.2 Priming of Reverse Transcription by Anti-sense RNA

This alternative (Fig. 2) invokes anti-sense RNA complementary to pre-mRNA as the primer. Variable lengths of double-stranded RNA originating in the Ei/MAR region and extending upstream towards the rearranged V(D)J would explain the production of mutated cDNA of variable length, thus accounting for the slope of the distribution of mutations terminating in the J-C intron. The same arguments as in the previous model account for the 5′ boundary of mutation. This model is not mutually exclusive with the DNA nicking model.

3 Gene Conversion in Chicken and Rabbit

In birds and rabbits the pre-antigen B cell repertoire is generated in gut-associated lymphoid tissues (the bursa of Fabricius in birds and the ilieal Peyer's patches of rabbit) by a process which has been described as gene conversion (McCormack and Thompson 1990a; Knight 1992). There is a substantial literature documenting various parameters of this process in the chicken. We will take the chicken light

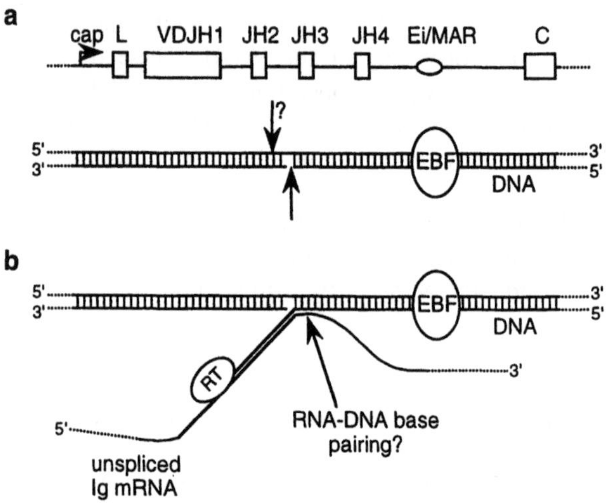

Fig. 1. **a** Nicking of a hypothetical murine IgH RNA-coding DNA strand after recognition of a specific sequence within the Ei/MAR region but before reverse transcription. Nicking can occur at any one of multiple recognition sequences (but only one is shown). Cleavage of the second strand may occur either before or after reverse transcription (indicated by *?*). **b** Reverse transcription of the unspliced Ig mRNA is initiated at the 3′ hydroxyl group exposed by the nick in the RNA-coding strand. The region of potential RNA–DNA heteroduplex formation is indicated. The cDNA then replaces the corresponding chromosomal DNA either before or after second strand synthesis, possibly via a DNA gap repair mechanism or via specific enzymic activities encoded by the mutatorsome or expressed by the cell. The relative positions of the cap site (*bent arrow*), leader (*L*), VDJ unrearranged JH elements, intronic enhancer/ matrix attachment region (*Ei/MAR)* and the constant gene (*C*) are shown. *EBF*, putative Ei/MAR binding factor; *RT*, reverse transcriptase

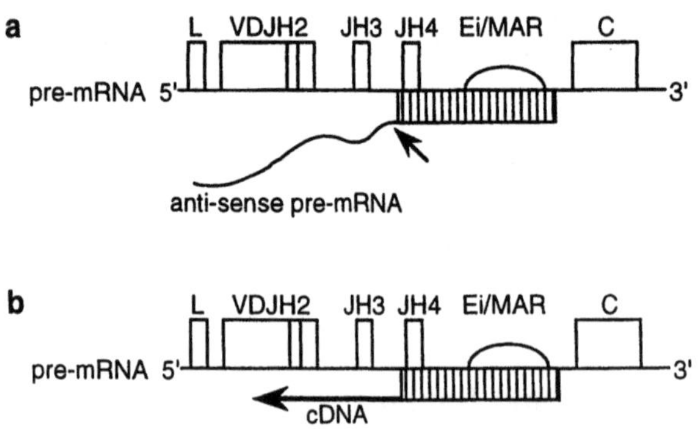

Fig. 2. **a** The amount of annealing between the pre-mRNA and the anti-sense pre-mRNA from the unrearranged or non-productive locus that occurs before the single-stranded RNA is nicked (indicated by *arrow*) determines the length of the anti-sense reverse transcriptase primer (shown in **b**). the relative positions of the leader (*L*), VDJ, unrearranged JH elements Ei/MAR and the constant gene (*C*) are shown

chain locus as a prototype to explore the possible molecular mechanisms involved. This locus contains only one fully functional V_L gene capable of rearrangement with a J_L segment, and subsequent expression as protein. The rearrangement process excises 1.8 kb of DNA between V_L and J_L, resulting in a circular episome joined by the V_L and J_L signal sequences. Upstream of the single functional V_L gene there is an array of 25 V_L pseudogenes crammed into 19 kb of DNA. A number of studies indicate that during the generation of the B cell repertoire in the bursa, after rearrangement of the single functional V_L and J_L genes, blocks of pseudogene V_L sequence are donated to the rearranged VJ. This process has been broadly termed gene conversion as it involves unidirectional donation of nucleotide sequence, but the precise mechanisms are unknown. To our knowledge all previous speculation assumes direct DNA-to-DNA donation.

We have approached the rationale underpinning this issue in the same way that we approached the somatic mutation issue in mouse and human B cells. In essence, we believe that such a mechanism *must* be limited to Ig loci because if it occurred in other multigene families during differentiation of the organism, severe biological problems would ensue. For example, somatic gene conversion in the MHC gene complex could result in autoimmunity and in disruption of cognate cell–cell interactions. For these fundamental reasons, we explored mechanisms analogous to somatic hypermutation of IgV genes based on the production of cDNA from an RNA template and exploiting the versatility and secondary structure potential of single-stranded RNA, together with DNA of the Ei/MAR region as a locus-specific localisation device.

In exploring this possibility we again invoked the necessity to protect the C region from deleterious sequence change and postulated the presence of an Ei/MAR region in the J-C intron of the chicken V_L locus, by analogy with such regions in the mouse V_H and V_κ loci. We then formulated models based upon long transcripts of RNA (commonly termed "sterile" transcripts) beginning upstream of the close-packed pseudogene array and proceeding through the functional V and J genes, the J-C intron and the C region. We then invoked secondary structure in Ei/MAR RNA to localise a molecular complex with reverse transcriptase activity analogous to the "mutatorsome" postulated above. Originally, we considered the long RNA transcript to behave like a ball of string with random loops protruding from it which provided an RNA template sequence for the production of cDNA by the reverse transcriptase, thus allowing production of cDNA based on pseudogene sequences present in some loops. These sequences could undergo homologous recombination with the rearranged V(J) chromosomal DNA. However, available data suggests non-random processes in the sequence donation process. For example, in the V_L locus of the chicken, eight out of the 25 pseudogenes are orientated in the opposite direction to the remaining 17 pseudogenes and the single functional V_L element. As a group, these eight pseudogenes donate sequence much more frequently than the other 17 pseudogenes (McCormack and Thompson 1990b). We then explored the possibility of the formation of non-random loops in the long RNA transcript. The simplest mechanism generating loops would involve substantial stretches of G-C or A-U base pairing. When sequences of the 25 V_L

pseudogenes are examined, two of them (V10 and V13) have poly-A sequences of 12 and 15 bases, respectively, immediately adjoining the downstream end of the coding sequence (REYNAUD et al. 1987). It would be extraordinary if such highly non-random sequence stretches had no functional significance. A search for a poly-T(U) sequence, which could base pair with either one of these poly-As and produce a loop in an RNA transcript, revealed a 14T stretch just over 1 kb downstream of the functional V_L gene and about 500 bases upstream of the functional J_L gene. Since the intervening sequence between the functional V_L and J_L genes is excised during DNA rearrangement, only RNA transcripts from the *unrearranged* locus would contain the poly-U sequence, which would enable a loop to be formed with the poly-A sequences abutting V10 or V13. Invoking a mandatory role for such a transcript would demand that in mature B lymphocytes of the chicken which have undergone gene conversion, there should be one unrearranged V locus. This seems to be the case, in marked contrast to B lymphocytes in mice in which both loci are often rearranged (review by MCCORMACK and THOMPSON 1990a).

The mechanism that we propose for the so-called gene conversion process is shown in Fig. 3. We postulate that in the long transcript of RNA from the unre-

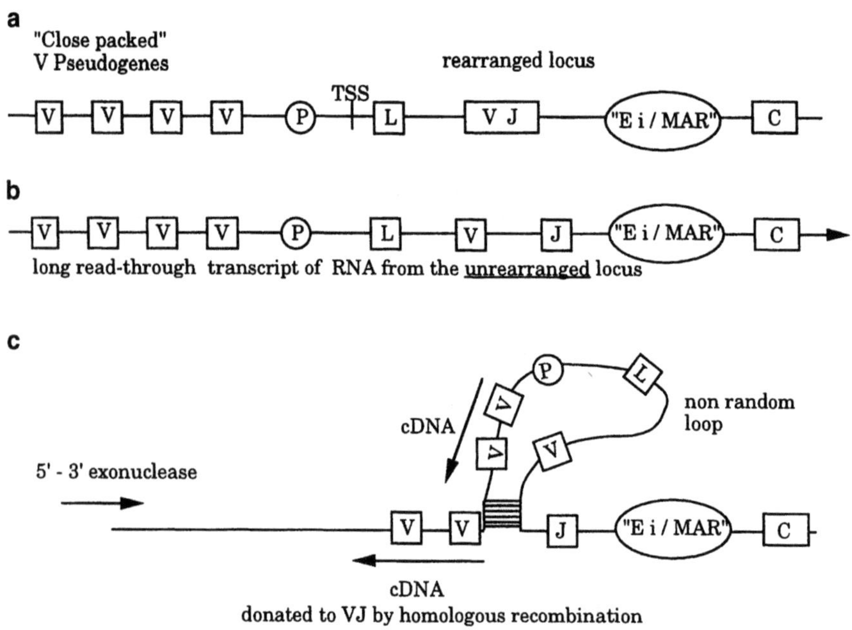

Fig. 3a–c. Gene conversion in the chicken V_L locus. The mechanism of donation of pseudogene sequence to rearranged VJ. a The chromosomal DNA in a rearranged locus. b A RNA transcript from the unrearranged locus which, in c, has formed a loop by base pairing as described in the text. We postulate that the RNA binds to the chromosomal DNA via locus-specific structures associated with the Ei/MAR region. Thus, pseudogene sequences are brought physically closer to the rearranged VJ locus. cDNA produced by reverse transcription of any of the RNA sequences within or outside the loop are potential donors by homologous recombination into the chromosomal DNA of the recipient VJ sequence. *TSS*, transcription start site

arranged locus, a loop forms by base pairing of the 14U sequence in the V_L-J_L intron with the 12A sequence abutting V10, or with the 15A sequence abutting V13. Since V13 is orientated in the same direction as the functional V_L and VJ sequences, the diagram showing the loop in the figure represents the case where the 15A sequence abutting V13 is used to base pair with the 14U sequence, so that V13 is immediately outside the loop and V10 is inside the loop. Since V10 is orientated in the opposite direction to the functional V_L and VJ sequences, if the 12A sequence abutting V10 is used to base pair and form the loop, V10 will be immediately within the loop. (Of course, in this anti-sense case, the 12As adjacent to V10 will actually be 12Us and the intron will contain a 14A stretch, but to avoid confusion we will stay with sense terminology.) We then postulate that a molecular complex analogous to the murine mutatorsome produces cDNA using RNA templates either inside or outside the loop. The function of the loop is to bring the pseudogene donor sequences in the RNA transcript physically closer to the Ei/MAR secondary structure to which the reverse transcriptase complex binds in a locus-specific manner and to protect RNA sequences within the loop from degradation by 5′ to 3′ exonuclease activity. A second specific binding event to the Ei/MAR region of chromosomal DNA of the rearranged locus would then bring the whole RNA-protein complex close to the recipient rearranged VJ DNA. Homologous recombination of cDNA derived from reverse transcription of templates in the RNA transcript from the unrearranged locus could then homologously recombine with the chromosomal VJ DNA of the rearranged locus. Clearly, such a process would require ordered regulation of transcript production and various competing processes would determine the nature of sequences donated and the rate of donation.

This model has striking correlations with published data and one major contradiction with published interpretation of the chicken gene conversion process. Firstly, it provides an explanation for the virtually universal presence of one unrearranged IgV locus in mature chicken B cells. Secondly, it accounts for the presence of point mutations in addition to apparent gene conversion in rearranged chicken IgV sequences (Reynaud et al. 1987; Arakawa et al. 1996), since the production of RNA and cDNA are error-prone copying processes, as invoked for somatic hypermutation in mouse and human. Thirdly, the location of the postulated loop in the long RNA transcript also correlates strikingly with published data in Fig. 4 of McCormack and Thompson (1990b) This figure shows the frequency of V pseudogene usage as gene conversion donors. Two points emerge from these data. Firstly, as a group, the eight pseudogenes with opposite orientation (anti-sense) to the functional V_L gene donate sequence more frequently than their counterparts orientated in the same direction (sense) as the functional V_L gene. Our interpretation of this result is that the production of the long RNA transcript in the anti-sense orientation is quantitatively greater than that in the sense orientation (because they will be independently regulated), thus providing quantitatively greater opportunity for donation from the anti-sense transcript. Secondly, as a group, pseudogenes located within the putative loop of RNA (V12 to V1) donate sequence more frequently than those outside the loop (V13 to V25). This is particularly striking for genes in the sense orientation. With our model this could be accounted for by the vulnerability of the single-stranded RNA template outside the loop to degradation by 5′ to 3′ exonuclease (see Fig. 3).

The apparent contradiction with the interpretation of chicken gene conversion data concerns whether or not donations are cis or trans. CARLSON et al. (1990) have interpreted their data to mean that donations always occur in cis. However, in our model it is clear that any gene located within our postulated RNA loop would be eligible as a donor. This includes the functional V_L sequence. Since this sequence is highly homologous in the leader and leader intron regions 5′ of the V_L coding sequence, we postulate that it almost invariably dominates donation because of that extensive homology in the 5′ region and because of the 5′ to 3′ direction of integration of cDNA, in concordance with the conclusion of MCCORMACK and THOMPSON (1990b). Additional smaller donations from the leaderless pseudogenes would be superimposed on the large donation from the functional V_L sequence. These postulates render our model compatible with the CARLSON et al. data.

Finally, this model unifies, at the level of the basic molecular mechanisms, somatic hypermutation and somatic gene conversion in B lymphocytes of higher vertebrates. Since species such as the rabbit employ both somatic mutation and gene conversion to generate a B cell repertoire, it seemed to us more probable that common molecular mechanisms are involved than otherwise.

Experimental testing of the chicken model would require manipulation of the loci with the goal of interfering with the postulated single-stranded nucleic acid intermediates. This could range from constructing transgenic chickens with both loci in rearranged form to more subtle intervention such as the removal of the 14T sequence in the V-J intron (GOODNOW, personal communication).

References

Arakawa H, Furusawa S, Ekino S, Yamagishi H (1996) Immunoglobulin gene hyperconversion ongoing in the chicken splenic germinal centers. EMBO J 15:2540–2546

Azuma T, Motoyama N, Fields LE, Loh DY (1993) Mutations of the chloramphenicol acetyl transferase transgene driven by the immunoglobulin promoter and intron enhancer. Int Immunol 5:121–130

Berek C, Berger A, Apel M (1991) Maturation of the immune response in germinal centers. Cell 67:1121–1129

Betz AG, Neuberger MS, Milstein C (1993) Discriminating intrinsic and antigen-selected mutational hotspots in immunoglobulin V genes. Immunol Today 14:404–411

Betz AG, Milstein C, Gonzalez-Fernandez A, Pannell R, Larson T, Neuberger MS (1994) Elements regulating somatic hypermutation of an immunoglobulin κ gene: critical role of the intron enhancer/matrix attachment region. Cell 77:239–248

Blackburn EH (1992) Telomerases. Annu Rev Biochem 61:113–129

Both GW, Taylor L, Pollard JW, Steele EJ (1990) Distribution of mutations around rearranged heavy-chain antibody variable-region genes. Mol Cell Biol 10:5187–5196

Carlson LM, McCormack WT, Postema CE, Humphries EH, Thompson CB (1990) Templated insertions in the rearranged chicken IgL V gene segment arise by intrachromosomal gene conversion. Genes Dev 4:536–547

Chen C, Roberts VA, Rittenberg MB (1992) Generation and analysis of random point mutations in an antibody CDR2 sequence: many mutated antibodies lose their ability to bind antigen. J Exp Med 176:855–866

Chen C, Roberts VA, Stevens S, Brown M, Stenzel-Poore MP, Rittenberg MB (1995) Enhancement and destruction of antibody function by somatic mutation: unequal occurrence is controlled by V gene combinatorial associations. EMBO J 14:2784–2794
Chien NC, Pollock RR, Desaymard C, Scharff MD (1988) Point mutations cause the somatic diversification of IgM and IgG2a anti-phosphorylcholine antibodies. J Exp Med 167:954–973
Giusti AM, Manser T (1993) Hypermutation is observed only in antibody H chain V region transgenes that have recombined with endogenous immunoglobulin H DNA: implications for the location of *cis*-acting elements required for somatic mutation. J Exp Med 177:797–809
Gonzalez-Fernandez A, Milstein C (1993) Analysis of somatic hypermutation in mouse Peyer's patches using immunoglobulin κ light-chain transgenes. Proc Natl Acad Sci USA 90:9862–9866
Hackett J, Stebbins C, Rogerson B, Davis MM, Storb U (1992) Analysis of a T cell receptor gene as a target of the somatic hypermutational mechanism. J Exp Med 176:225–231
Jacob J, Kelsoe G, Rajewsky K, Weiss U (1991) Intraclonal generation of antibody mutants in germinal centres. Nature 354:389–392
Knight KL (1992) Restricted V_H gene usage and generation of antibody diversity in rabbit. Annu Rev Immunol 10:593–616
Lebecque SG, Gearhart PJ (1990) Boundaries of somatic mutation in rearranged immunoglobulin genes: 5′ boundary is near the promoter, and 3′ boundary is approximately 1 kb from V-D-J gene. J Exp Med 172:1717–1727
Luan DD, Korman MH, Jakubczak JL, Eickbush TH (1993) Reverse transcription of R2Bm RNA is primed by a nick at the chromosomal target site: a mechanism for non-LTR retrotransposition. Cell 72:595–605
Manser T (1990) The efficiency of antibody affinity maturation: can the rate of B-cell division be limiting? Immunol Today 11:305–308
McCormack WT, Thompson CB (1990a) Somatic diversification of the chicken immunoglobulin light-chain gene. Adv Immunol 48:41–67
McCormack WT, Thompson CB (1990b) Chicken IgL variable gene conversion display pseudogene donor preference and 5′ to 3′ polarity. Genes Dev 4:548–558
Motoyama N, Okada H, Azuma T (1991) Somatic mutation in constant regions of mouse ll light chains. Proc Natl Acad Sci USA 88:7933–7937
Peters A, Storb U (1996) Somatic hypermutation of immunoglobulin genes is linked to transcription initiation. Immunity 4:57–65
Rada C, Gonzalez-Fernandez A, Jarvis, JM, Milstein C (1994) The 5′ boundary of somatic hypermutation in a V_κ gene is in the leader intron. Eur J Immunol 24:1453–1457
Reynaud C-A, Anquez V, Grimal H, Weill J-C (1987) A hyperconversion mechanism generates the chicken light chain preimmune repertoire. Cell 48:379–388
Rogerson BJ (1994) Mapping the upstream boundary of somatic mutations in rearranged immunoglobulin transgenes and endogenous genes. Mol Immunol 31:83–98
Rothenfluh HS, Taylor L, Bothwell ALM, Both GW, Steele EJ (1993) Somatic hypermutation in 5′ flanking regions of heavy chain antibody variable genes. Eur J Immunol 23:2152–2159
Sohn J, Gerstein RM, Hsieh C-L, Lemer M, Selsing E (1993) Somatic hypermutation of an immunoglobulin u heavy chain transgene. J Exp Med 177:493–504
Steele EJ (1991) (ed) Somatic hypermutation in V-regions. CRC Press, Boca Raton, Florida
Steele EJ, Pollard JW (1987) Hypothesis: somatic hypermutation by gene conversion via the error prone DNA → RNA → DNA information loop. Mol Immunol 24:667–673
Steele EJ, Rothenfluh HS, Both GW (1992) Defining the nucleic acid substrate for somatic hypermutation. Immunol Cell Biol 70:129–144
Steele EJ, Rothenfluh HS, Blanden RV (1997) Mechanism of antigen-driven somatic hypermutation of rearranged immunoglobulin V(D)J genes in the mouse. Immunol Cell Biol 75:82–95
Tonegawa S (1983) Somatic generation of antibody diversity. Nature 302:575–581
Umar A, Gearhart PJ (1995) Reciprocal homologous recombination in or near antibody VDJ. Eur J Immunol 25:2392–2400
Umar A, Schweitzer PA, Levy NS, Gearhart JD, Gearhart PJ (1991) Mutation in a reporter gene depends on proximity to and transcription of immunoglobulin variable transgenes. Proc Natl Acad Sci USA 88:4902–4906
Weber JS, Berry J, Manser T, Claflin JL (1991) Position of the rearranged V_κ and its 5′ flanking sequences determines the location of somatic mutations in the J_κ locus. J Immunol 146:3652–3655
Xu B, Selsing E (1994) Analysis of sequence transfers resembling gene conversion in a mouse antibody transgene. Science 265:1590–1593

Yang J, Zimmerly S, Perlman PS, Lambowitz AM (1996) Efficient integration of an intron RNA into double-stranded DNA by reverse splicing. Nature 381:332–335

Yang X, Stedra J, Cerny J (1996) Relative contribution of T and B cells to hypermutation and selection of the antibody repertoire in germinal centers of aged mice. J Exp Med 183:959–970

Yelamos J, Klix N, Goyenechea B, Lozano F, Chui YL, Gonzalez-Fernandez A, Pannell R, Neuberger MS, Milstein C (1995) Targeting of non-Ig sequences in place of the V segment by somatic hypermutation. Nature 376:225–229

Zimmerly S, Guo H, Perlman PS, Lambowitz AM (1995) Group II intron mobility occurs by target DNA-primed reverse transcription. Cell 82:545–554

Characteristics of Somatic Hypermutation of Human Immunoglobulin Genes

R.A. INSEL and W.S. VARADE

1 Introduction

Affinity maturation of antibodies after immunization is a result of hypermutation of the variable region of immunoglobulin genes and an antigen selection process that preserves those B lymphocytes with mutated surface immunoglobulin (Ig) expressing an increased affinity for the immunogen (FRENCH et al. 1989; KOCKS and RAJEWSKY 1989; WAGNER and NEUBERGER 1996). The surviving B cells then give rise to memory B cells and antibody secreting cells. Hypermutation is the predominant mechanism for diversification of the secondary antibody response. The hypermutation process is not only lineage-specific but is site- and stage-specific. Hypermutation of Ig genes is prominent in B cells in the germinal centers of secondary lymphoid organs from approximately 1–3 weeks after immunization (LEVY et al. 1989; BEREK et al. 1991; JACOB et al. 1991; MACLENNAN 1994; PASCUAL et al. 1994). The mutations are primarily single point substitutions that are targeted to 2 kb of DNA of rearranged heavy and light chain V gene segments with a sharp upstream boundary in the middle of the leader intron and an imprecise downstream boundary in the J-C intron that are introduced at approximately 10^{-3}–10^{-4} mutations per base pair per generation, which is the highest rate of mutation observed

University of Rochester School of Medicine, Department of Pediatrics, 601 Elmwood Avenue, P.O. Box 777, Rochester, NY 14642, USA
Supported by National Institutes of Health Grants AI37123 and 1K08DK02316

in the eukaryotic genome (Kocks and Rajewsky 1989; Lebeque and Gearhart 1990; Wagner and Neuberger 1996).

The mechanism of hypermutation of Ig genes is not well understood at a cellular or molecular level. We have investigated characteristics of the mutation process of human V genes to gain further insights into its mechanism and to identify those characteristics of the mutation process that are common between humans and other species. In these studies, we have analyzed translationally silent and, thus, antigen non-selected mutations of human V genes to gain an understanding of intrinsic properties of the mutation process (Varade et al. 1993; Varade and Insel 1993; Insel et al. 1994; Insel and Varade 1994). This analysis has been expanded to analyze the inherent properties of all mutations, including potential replacement and silent mutations in the absence of antigen selection by analysis of the mutations of human nonproductive V gene rearrangements, which function naturally as "passenger" genes (Varade et al. 1997). To determine the rate of mutation of human V genes, we have developed a system to induce, recover, and characterize mutations that occur in a germline human V gene after in vivo immunization (Insel et al. 1997). The results of these studies have provided insights into the mutation process.

2 Inherent Properties of Somatic Hypermutation of Human Immunoglobulin Genes

A large group of silent mutations of the human V_H6 gene, a single member human V_H family that has not been associated with polymorphisms, which makes this gene an excellent substrate to unambiguously identify mutations, were analyzed (Varade et al. 1993; Varade and Insel 1993; Insel et al. 1994; Insel and Varade 1994). The human V_H6 repertoire is quite diverse with no evidence that it is recruited by only a single antigen. IgM, IgG, and IgA V_H6 Ig gene rearrangements were analyzed from human spleens surgically removed for hematologic indications. Single rearrangements generated multiple unique clones with evidence of intraclonal diversification with unique point mutations and isotype switching, which was consistent with their derivation from germinal centers (Insel and Varade 1994).

The overall mutation frequency of the productive V_H6 rearrangements analyzed was 4.5% (767 of 17 068 bases sequenced) and was higher for IgG (5.4%) and IgA (5.2%) than IgM (3.2%) expressing clones. The distribution of mutations across the V_H6 gene showed that the majority were concentrated in the complementarity-determining regions (CDR) (CDR1, 9.2%; CDR2, 6.6%) vs the framework regions (FR) (3.4%). The replacement to silent mutation ratio (R/S) across the V_H6 gene showed a higher ratio in CDR1 (4.6) and CDR2 (2.7) than in the FR (1.6). The R/S ratio for the total V gene and for the CDR of IgM clones (3.0, 5.1, respectively) was higher than IgG (1.8, 2.3, respectively) or IgA clones (2.0, 3.8, respectively). The high frequency of mutation and R/S ratio in the CDRs vs the FR

(3.2 vs 1.6), were consistent with antigen selection of these rearrangements. The higher frequency of mutations and the lower R/S ratios in the IgG and IgA clones suggest that repetitive antigenic stimulation may have induced new rounds of somatic mutation and resulted in isotype switching. Excess replacement mutations, which may be deleterious for preservation of Ig function (CHEN et al. 1992), may have been selected against with accumulation of silent mutations, leading to lower R/S ratios. The highly mutated IgM clones, which were probably derived from IgD negative memory splenic B cells, have a potential to develop a more appropriate effector function after repetitive antigenic challenge that may elicit isotype switching and new rounds of mutation. This concept is consistent with the observation that isotype switching occurs later than somatic mutation in germinal centers (LIU et al. 1996).

To analyze the inherent properties of somatic hypermutation, the silent mutations, which are not antigen selected, of these productive rearrangements were analyzed for the type and targeting of mutations (INSEL and VARADE 1994). An expected frequency of replacement and silent mutations, transitions, and transversions for each base position for the germline V_H6 gene was calculated based on the germline V_H6 sequence assuming randomness without negative or positive selection, except when a stop codon was generated, which was discounted as an event.

To extend this analysis more generally across the V region and not be dependent on the generation of silent mutations at a particular nucleotide, we analyzed all the mutations that occurred in a group of nonproductive rearrangements of human V_H6 rearrangements (VARADE et al. 1997). Nonproductive rearrangements were enriched from genomic splenic DNA by subtractive hybridization with cDNA as the driver for subtraction. Mutations in those rearrangements that were nonproductive based on a loss of translational reading frame due to a CDR3 frameshift (nine rearrangements) or generation of a junctional stop codon (one rearrangement) were analyzed for bias. The frequency of mutations in these nonproductive rearrangements was 3.0%, which was similar to productive rearrangements, which indicates mutations are targeted equally to productive and nonproductive rearrangements, as described in the mouse (ROES et al. 1989; MOTOYAMA et al. 1994).

2.1 The CDR of Some Human V Genes are Inherently Susceptible to Mutation

A total of 167 unique silent mutations in the 38 V_H6 productive rearrangements were analyzed. First, there was bias in the distribution of these silent mutations across the V_H6 gene. There were 54 silent mutations in the CDR compared to the 34.7 expected and, similarly, there were 113 FR mutations compared to the 132.3 expected ($p < 0.001$). The bias was primarily due to an increased frequency of silent mutations in both CDR1 and CDR2 and a decreased frequency of silent mutations in FR1. This excess of silent mutations in the CDR could possibly be explained if there was co-selection of silent mutations with antigenic selection of replacement mutations in the CDR. Some findings suggest that, in fact, mutations of human V genes may occur in pairs, clusters, or blocks (KLEIN et al. 1993; see also Sect. 2.3 below). Alternatively, the CDR of the V_H6 gene may be intrinsically more mutagenic. To analyze for this possibility, the 83 mutations of the nonproductive V_H6 rearrangements were analyzed. There was also evidence of targeting of mutations to the CDR in the nonproductive rearrangements (40% observed vs 27% expected) compared to the FR (60% observed vs 73% expected) ($p < 0.01$). Thus, CDR1 and CDR2 of human V_H6 are inherently mutagenic. Not only were a greater number of mutations targeted to the CDR of V_H6, but when mutations in the CDR of the nonproductive rearrangements occurred, a preferential bias for replacement over silent mutations, compared to the expected based on the germline sequence, was observed. This was not observed in the FR. Some of the properties of the sequences of the CDR of V_H6 that leads to this inherent susceptibility to mutation are described below.

In a similar fashion, murine V_L passenger transgenes display an excess frequency of CDR1 mutations (BETZ et al. 1993a). It has been suggested that CDR sequences may have been selected and conserved with evolution because of their inherent mutagenicity (BETZ et al. 1993b; WAGNER and NEUBERGER 1996). The targeting of CDR mutations in V_H6 could represent an evolutionary selection of mutable CDRs for this nonpolymorphic gene whose overall structure has been highly conserved in primate evolution (MEEK et al. 1991). Conceivably the V_H6 gene encodes antibodies to an important pathogen in primates, which, as yet, has not been identified.

2.2 Preference for Transitions and C→G and G→C Transversions With Mutation

A bias for transitions vs transversions was observed for the silent mutations of the productive and for all mutations of the nonproductive rearrangements. The expected transition to transversion ratio for all potential silent mutations in germline V_H6 was 0.84 and, yet, there were 115 transitions and 52 transversions, giving a ratio of 2.2 ($p < 0.001$). Similarly, in the nonproductive rearrangements, the ratio of transitions to transversions was 0.80 compared to the expected ratio of 0.5 ($p < 0.05$). An excess of transitions was observed for the silent mutations of the productive rearrangements for each of the four bases and for A and C nucleotides for the nonproductive rearrangements. Similarly, the 94 unique silent mutations in the productive rearrangements at the third position in codons in which mutation at the third position generates a silent mutation regardless of the substitution, yielded almost an equal number of transitions and transversions (44 transitions, 50 transversions), compared to the expected one-third transition and two-thirds transversion frequency that would occur randomly at this position ($p < 0.05$). This finding of an excess of transitions to transversions is similar to that observed in mutations of mouse Ig gene sequences selected for antigenic specificity, Ig and non-Ig passenger transgenes in the mouse, V regions in the sheep primary antigen-independent antibody repertoire, and V regions in the frog (GOLDING et al. 1987; WILSON et al. 1992; BETZ et al. 1993; JACOB et al. 1993; REYNAUD et al. 1995).

Although transversions were disfavored compared to transitions, there was a distinct bias for specific types of transversions. There was bias for C → G over C → A transversions (25 vs 6), and for G → C over G → T transversions (32 vs nine) when silent transversions in the third position of codons that had an equal chance of giving rise to one of two replacement mutations, silent transversions, and all transversions in the nonproductive rearrangements were analyzed. This bias for C → G and G → C, transversions is not explained by codon usage in the human genome and may reflect a fundamental conserved feature of the mutation process. A similar bias has been recently described in mutations of the non-coding V regions of a human follicular lymphoma cell line and of non-coding murine J region genes (WU and KAARTINEN 1996), and possibly of some passenger murine transgenes (BETZ et al. 1993a; YÉLAMOS et al. 1995). A C → G vs a C → A transversion preference was also observed for a His to Gln substitution in murine antibodies to 2-phenyloxazolone (BETZ et al. 1993a).

It is of interest that mutations of Ig genes in the frog preferentially target G-C base pairs (WILSON et al. 1992). Although G-C base pair targeting of mutations is not observed in murine V genes, WABL and colleagues observed a targeting of G-C base pairs in the reversion of a stop codon in a mouse IgM VDJ rearrangement that had been transfected into the murine pre-B cell line 18–81 (BACHL and WABL 1996). WABL suggested that the targeting of G-C base pairs may represent a primary mutator whose effects are obscured by antigen selection acting on productive rearrangements or one of several mutators that normally acts to diversify Ig rearrangements. Of interest to the transversion bias described above, the reversion of the stop codon represented a preferential bias for G → C over G → T transversions at a 7:1 ratio. If such a unique mutator exists, then this G-C mutator has a marked bias for transversions to C and G.

2.3 Nucleotide Targeting of the Mutation Process

There was preferential targeting of mutations to single, di-, and trinucleotides in human V genes. There was an increase of silent mutations involving adenine residues and a decrease of mutations involving thymine residues on the coding strand compared to the expected for the silent mutations of the productive V_H6 rearrangements (INSEL and VARADE 1994). This bias against mutation of thymine is similar to what was described in murine germinal centers and for murine passenger V_L transgenes (GOLDING et al. 1987; BETZ et al. 1993; JACOB et al. 1993). These findings suggest that there is a bias for targeting of purines with the mutations exhibiting strand polarity or preferential targeting of the mutation process to one of the two strands.

Strand polarity, however, has not been consistently observed with somatic hypermutation. We did not observe strand polarity or disfavoring of mutations of thymine nucleotides in the mutations of the V_H6 nonproductive rearrangements

(Varade et al. 1997), nor in silent mutations of V_H5 rearrangements (unpublished observation). Strand polarity with an A to T preference has been found for mutations of some non-Ig murine passenger transgenes (bacterial *neo*) but not in others (bacterial *gpt*, human b-globin) (Y'lamos et al. 1995).

The status of strand polarity being an inherent property of the mutation process is still debatable. There are potential mechanisms that could lead to polarization of the mutation process. If mutation is activated during replication, which is an asymmetric process for the leading and lagging strands of double-stranded DNA, or during transcription and/or repair (Manser 1990; Rogerson et al. 1991; Umar et al. 1991; Betz et al. 1993b; Peters and Storb 1996), then strand discrimination for mutation could occur. If such is occurring, however, then its detection must be highly dependent on the substrate studied and any intrinsic hot spots for mutation in this substrate. Definitive experiments to prove strand polarity of mutation have not been performed to date.

When mutation frequencies of di- and trinucleotides were calculated and normalized to the specific di- and trinucleotide composition of the germline V_H6 sequence in all reading frames, a bias in decreasing order for favoring of mutations of TA, AT, TT and GC dinucleotides and GTA, TAC, CTA, TAG, AGC and ATA trinucleotides was observed. Bias was seen against mutations of CG, TC, GA, and CC. The germline dinucleotides appeared to be more important in targeting mutation than the trinucleotide sequences as the trinucleotide targets of mutation in the V_H6 nonproductive rearrangements reflected the presence of the preferred dinucleotide. TA, the most mutated dinucleotide, was contained in five of the top six highly mutable trinucleotides. There was augmentation of the bias observed for TA alone only in the trinucleotide GTA and not the other TA-containing trinucleotides. AT and GC dinucleotides were contained in four and two of the top thirteen trinucleotides, respectively. The dinucleotides most favored – TA, AT, and GC – are palindromic and thus can not be used to address the issue of strand polarity of the mutation process.

A high frequency of mutations at GC, TA, AC, AA, AT, and AG dinucleotides, in decreasing order, and in TA-containing trinucleotides was observed for murine V regions (Smith et al. 1996). Thus, it appears that TA, AT and GC dinucleotides are preferred targets of somatic mutation in the mouse as well as in humans. It is of interest that this TA targeting contrasts with the stop codon revertants in transfectants into murine pre-B cell lines in which mutation targeting to TAG occurs only at the G residue (Bachl and Wabl 1996), as described above.

The consensus RGYW, where R is A or G, Y is C or T, W is A or T (AGTA, AGTT, AGCT, AGCA, GGTA, GGTT, GGCA, GGCT) and TAA have been identified as hot spots for mutations (Rogozin and Kolchanov 1992), and CAGCT and AAGTT, which contain the RGYW consensus, were identified by Betz and colleagues as hot spots for mutation (Betz et al. 1993a, b). The RGYW consensus frequently contains the mutagenic dinucleotides described above and some have the potential for stem loop formation, as described below. The CDR of the human V_H6 gene are not enriched in RGYW sequences compared to the FR, but there is a disproportionate increased number of TA, AT, and TT which are favored for mutation and a decrease in GA, CT, and TC which are disfavored for mutation in the CDR as compared to the FR. This dinucleotide distribution bias may in part explain the increased frequency of mutations targeted to the CDR of this gene.

A clustering of mutations in the human V_H6 genes was also observed. Of the 83 silent and replacement mutations in the nonproductive V_H6 rearrangements, 22 or 26% occurred in adjacent nucleotides. Similarly, 16% of mutations in the productive rearrangements were clustered. Clustering of mutations or so-called "block" mutations has been observed in mutations of human V_κ genes (KLEIN et al. 1993). This raises the question whether mutations are not introduced as single point mutations with each or every other round of DNA replication but as multiple mutations in a single round with preferential targeting to hot spots. This possibility has been raised previously (YÉLAMOS et al. 1995).

Both local primary sequence motifs and higher order structural motifs may prove critical for targeting of mutation. There were some nucleotide positions that were heavily targeted for mutation that expressed inverted repeats, or palindromes, which have been suggested as sites susceptible to mutation (GOLDING et al. 1987; KOLCHANOV et al. 1987). There was potential stem-loop formation at codons 60–62 CAG > TAT < CTG in CDR2 of V_H6 with the tip (TAT) encoding 15 silent and 12 replacement mutations and the bordering G and C another 12 replacement mutations in the productive rearrangements. Codons 80–83 in FR3 TGCAG-CTGAA also showed potential stem-loop formation with the tip (GC) having nine silent and seven replacement mutations. In the nonproductive rearrangements, a hotspot for mutation was observed at the inverted repeat in FR3 codons 89–91 GTA > T < TAC with over 10% of the mutations of the nonproductive rearrangements and about 5% of the mutations in the productive rearrangements occurring in these seven nucleotides. In total, about one-fourth of the mutations of the nonproductive rearrangements occurred at these three inverted repeats. The mutagenic GTA trinucleotide and the TA and AT dinucleotide described above are prominently expressed in several of these inverted repeats. Both CDR1, at codons 27–31, and CDR2 at codons 60–62 of the V_H6 genes displayed inverted repeats, which may contribute to their inherent mutagenicity. Inverted repeats were not found, however, at all hot spots.

2.4 Mechanisms of Somatic Hypermutation

The mechanism of somatic hypermutation is not known. Among mechanisms proposed are an error-prone repair process, a transcription-coupled repair defect, gene conversion, unique polymerases, and a lagging strand DNA synthesis defect (BRENNER and MILSTEIN 1966; MANSER 1990; ROGERSON et al. 1991; UMAR et al. 1991; PETERS and STORB 1996). There is the possibility of induction in germinal centers of a unique polymerase or alteration of fidelity of an ubiquitous polymerase that leads to hypermutation.

Of interest is our recent observation that telomerase expression is markedly upregulated in human germinal center B cells (HU et al. 1997). Telomerase is a ribonucleoprotein polymerase that repairs the 5′ end of the lagging strand that shortens with replication and maintains telomere length. Shortening of the telomere of B lymphocytes with repetitive proliferation in the germinal center could limit the life span of a memory B cell. We found that as human tonsil B cells evolve from IgD^{+}, $CD38^{-}$ naive B cells to become IgD^{-}, $CD38^{+}$ germinal center B cells, there is a 100–1000-fold increase in telomerase activity. The upregulation of telomerase in germinal center B cells may act to preserve telomere length in memory B cells and, thus, their longevity. Memory B cells could be activated in vitro through sIg or CD40 to upregulate telomerase expression (Hu et al. 1997). The mechanism of induction of telomerase is currently being investigated, but the finding suggests a precedent for expressing novel polymerases or alteration of fidelity of a polymerase at this site.

3 Induction of Human V Gene Mutations and Estimation of Rates of Mutation

The mutation process in murine Ig genes has been studied by exploiting well-defined, restricted, dominant antibody responses to haptens. No similar system exists in man for inducing and characterizing the rate and nature of mutations after immunization. We have developed an approach to induce mutations in human Ig V genes.

The principal virulence determinant of the bacteria *Haemophilus influenzae* b (Hib) is its capsular polysaccharide (CP), a repetitive polymer of ribosyl-ribitol-phosphate. The antibody response to the HibCP is one of the most extensively studied human antibody responses to date because of restricted diversity of the antibody response (Insel et al. 1992) and the development of Hib vaccines (Ellis and Granoff 1994). The majority of adults immunized with HibCP vaccines generate a predominant V_k II gene-A2 encoded response. The CDR3 region of A2 encoded antibody, which arises from V_L-J_L joining, is unique in that it is one amino acid longer than the typical nine-amino acid V_k CDR3 with an invariant arginine at the V_k–J_k junction.

We have cloned and sequenced the A2 repertoire in peripheral blood lymphocytes (PBL) prior to and after immunization with HibCP vaccines and taken advantage of A2 predominance and the use of the prototypic (10aa CDR3 with arginine) A2 sequence in the HibCP antibody response, the relative lack of polymorphism of the A2 gene in man, and our observation that a high proportion of circulating human PBL collected at 7–8 days after immunization with HibCP vaccines secretes antibody to the CP (Insel et al. 1997; Munoz and Insel 1987). Single rearrangements with the prototypic ten-amino acid CDR3 with an arginine (CGT) at the V_k–J_k junction that arose by presumed N region addition were isolated. These rearrangements gave rise to up to 19 unique clones with evidence of extensive diversification. Some of these rearrangements were recruited into the repertoire as a germline sequence by immunization as evidenced by failure to detect the rearrangement in preimmunization libraries from the immunized subject, the isolation of germline clones arising from the same rearrangement, and the finding that no one single mutation was shared by all the clones. Both shared and unique somatic mutations were apparent in the clones. The frequency of unique mutations was almost 1% at day 7 after immunization with an average of 2.8 unique mutations per clone. This type of diversification was similar to that observed in V gene sequences isolated from murine germinal centers (BEREK et al. 1991; JACOB et al. 1991, 1993). Based on a cell cycle time of approximately 10 h over a 7-day period (JACOB et al. 1991, 1993), it is possible to estimate a mutation rate of 10^{-3}–10^{-4}/bp per generation, which is consistent with estimates of the rate of mutation of murine V genes (KOCKS and RAJEWSKY 1989; FRENCH et al. 1989; BEREK and MILSTEIN 1987; MCKEAN et al. 1984; JACOB et al. 1993).

An estimate of mutation rate may be affected by ongoing antigenic selection of mutations. The R/S ratios for the unique mutations showed a lower R/S in the CDR than expected, and in some immunized subjects a lower R/S in the CDR than the FR. This indicates preservation of the germline CDR sequence, suggesting that the germline A2 sequence, which has been conserved in the human population, may be ideal for antigen binding. If anything, negative selection of mutation has occurred. Counterselection of CDR replacement mutations in this antibody response must be generating a high degree of cell loss during diversification of capsular polysaccharide-specific B cells through loss or decrease in antibody binding due to replacement mutations in either the CDR or in the FR.

Analysis of the genealogical relationships of the mutations indicated that the B cells with these A2 rearrangements were recruited by immunization and then sustained multiple cell divisions prior to undergoing mutation. There were rare memory B cells expressing the A2 rearrangements isolated at day 7 after immunization that were detectable months after immunization in the circulation. Both mutated and germline A2 sequences were isolated, confirming that negative pressure can preserve a germline sequence. This finding also suggests that the estimated mutation rate has not been excessively biased by positive selection of mutations. If anything, the CDR of A2-encoded antibody behaved as a FR with this negative antigen selection pressure. There may be inherent limitations to increased affinity of antibodies to polysaccharides by somatic mutation from enthalpy-entropy effects.

The inherent R/S of the CDR of the A2 germline gene is not different from the mean R/S of a diverse collection of human V_L genes. Also, the CDR of the germline A2 gene does contain the dinucleotide (VARADE et al. 1997; SMITH et al. 1996) and other motifs (ROGOZIN and KOLCHANOV 1992; BETZ et al. 1993a, b) that have been defined as major intrinsic hotspots of somatic mutation. The A2 gene appears to have been conserved in the population presumably because it encodes an optimal light chain sequence for protective antibody to an important human pathogen.

4 Conclusions

Fundamental conserved features of somatic hypermutation include a bias for transitions and G↔C transversions of G-C base pairs and targeting of TA, AT, and GC, dinucleotides. A targeting of stem loops appears also to be a prominent feature of mutation. The clustering of mutations raises the question of whether multiple mutations are introduced per round rather than just single point mutations. Strand polarity remains an intriguing but unproven property of the mutation process. The CDR of the human V_H6 gene are inherently mutagenic presumably because of their high content of mutagenic dinucleotides and inverted repeats with potential stem loops. Some human V gene sequences may be conserved because of their inherent mutageneity. Not all V genes with the potential for mutation are conserved because of their mutation potential, however. The V_k-A2 light chain, which encodes antibody to an important human pathogen, has the potential for mutation but negative selection preserves its germline sequence. A mutation rate of 10^{-3}–10^{-4}/bp per generation was observed after immunization in this human V gene. There was evidence that proliferation preceded mutation and that mutations were introduced stepwise with the mutation process either ceasing in a stochastic manner or being reactivated in some clones. Models of somatic hypermutation will need to take account of these findings. A thorough understanding of the mechanisms of mutation of Ig V genes will, however, require identification of the enzymes that mediate the somatic mutation process.

Acknowledgements. Supported by National Institutes of Health grant numbers AI37123 and 1K08DK02316.

References

Bachl J, Wabl M (1996) An immunoglobulin mutator that targets G -C pairs. Proc Natl Acad Sci USA 93:851–855

Berek C, Berger A, Apel M (1991) Maturation of the immune response in germinal centers. Cell 67: 1121–1129

Betz AG, Rada C, Pannell R, Milstein C, Neuberger MS (1993a) Passenger transgenes reveal intrinsic specificity of the antibody hypermutation mechanism: clustering, polarity, and specific hot spots. Proc Natl Acad Sci USA 90:2385–2388

Betz AG, Neuberger MS, Milstein C (1993b) Discriminating intrinsic and antigen-selected mutational hotspots in immunoglobulin V genes. Immunol Today 14:405–411

Brenner S, Milstein C (1966) Origin of antibody variation. Nature 211:242–243

Chen C, Roberts VA, Rittenberg MB (1992) Generation and analysis of random point mutations in an antibody CDR2 sequence: many mutated antibodies lose their ability to bind antigen. J Exp Med 176:855–866

Ellis RW, Granoff DM (1994) Development and clinical uses of *Haemophilus* b conjugate vaccines. Dekker, New York, pp 19–35

French DL, Laskov R, Scharff MD (1989) The role of somatic hypermutation in the generation of antibody diversity. Science 244:1152–1157

Golding GB, Gearhart PJ, Glickman BW (1987) Patterns of somatic mutations in immunoglobulin variable genes. Genetics 115:169–177

Hu B, Lee S, Marin E, Ryan D, Insel R (1997) Telomerase is upregulated in human germinal center B cells in vivo and can be re-expressed in memory B cells activated in vitro. J Immunol 159:1068–1071

Insel RA, Varade WS (1994) Bias in somatic hypermutation of human V_H genes. Int Immunol 6: 1437–1443

Insel RA, Adderson EE, Carroll WL (1992) The repertoire of human antibody to the *Haemophilus influenzae* type b capsular polysaccharide. Int Rev Immunol 9:25–42

Insel RA, Marin E, Varade WS (1994) Human splenic IgM immunoglobulin transcripts are mutated at high frequency. Mol Immunol 31:383–392

Insel RA, Marin E, Chu Y-W (to be published) Immunization induced diversification and mutation of a human antibody repertoire

Jacob J, Kelsoe G, Rajewsky K, Weiss U (1991) Intraclonal generation of antibody mutants in germinal centres. Nature 354:389–392

Jacob J, Przylepa J, Miller C, Kelsoe G (1993) In situ studies of the primary immune response to (4-hydroxy-3-nitrophenyl)acetyl. III. The kinetics of V region mutation and selection in germinal center B cells. J Exp Med 178:1293–1307

Klein R, Jaenichen R, Zachau HG (1993) Expressed human immunoglobulin genes and their hypermutation. Eur J Immunol 23:3248–3271

Kocks C, Rajewsky K (1989) Stable expression and somatic hypermutation of antibody V regions in B cell developmental pathways. Annu Rev Immunol 7:537–559

Kolchanov NA, Solovyov VV, Rogozin IB (1987) Peculiarities of immunoglobulin gene structures as a basis for somatic mutation emergence. FEBS Lett 214:87–90

Lebecque SG, Gearhart PJ (1990) Boundaries of somatic mutation in rearrangement immunoglobulin genes: 5′ boundary is near the promoter, and 3′ boundary is 1 kb from V(D)J gene. J Exp Med 172:1717–1727

Levy NS, Malipiero UV, Lebecque SG, Gearhart PJ (1989) Early onset of somatic mutation in immunoglobulin V_H genes during the primary immune response. J Exp Med 169:2007–2019

Liu YJ, Malisan F, de Bouteiller O, Guret C, Lebecque S, Banchereau J, Mills FC, Max EE, Martinez-Valdez H (1996) Within germinal centers, isotype switching of immunoglobulin genes occurs after the onset of somatic mutation. Immunity 4:241–250

MacLennan ICM (1994) Germinal centers. Annu Rev Immunol 12:117–139

Manser T (1990) The efficiency of antibody affinity maturation: can the rate of B-cell division be limiting? Immunol Today 11:305–307

McKean D, Huppi K, Bell M, Staudt L, Gerhard W, Weigert M (1984) Generation of antibody diversity in the immune response of BALB/c mice to influenza virus hemagglutinin. Proc Natl Acad Sci USA 81:3180–3184

Meek K, Eversole T, Capra JD (1991) Conservation of the most J_H proximal Ig V_H gene segment (V_HVI) throughout primate evolution. J Immunol 146:2434–2438

Motoyama N, Miwa T, Suzuki Y, Okada H, Azumal (1994) Comparison of somatic mutation frequency among immunoglobulin genes. J Exp Med 179:395–403

Munoz JL, Insel RA (1987) In vitro human antibody production to the *Haemophilus influenzae* type b capsular polysaccharide. J Immunol 139:2026–2031

Pascual V, Liu YJ, Magalski A, de Bouteiller O, Banchereau J, Capra JD (1994) Analysis of somatic mutation in five B cell subsets of human tonsil. J Exp Med 180:329–339

Peters A, Storb U (1996) Somatic hypermutation of immunoglobulin genes is linked to transcription initiation. Immunity 4:57–65

Reynaud CA, Garcia C, Hein WR, Weill J-C (1995) Hypermutation generating the sheep immunoglobulin repertoire is an antigen-independent process. Cell 80:115–125

Roes J, Huppi K, Rajewsky K, Sablitzky F (1989) V gene rearrangement is required to fully activate the hypermutation mechanism in B cells. J Immunol 142:1022–1026

Rogerson B, Hackett J, Peters A, Haasch D, Storb U (1991) Mutation pattern of immunoglobulin transgenes is compatible with a model of somatic hypermutation in which targeting of the mutator is linked to the direction of DNA replication. EMBO J 10:4331–4341

Rogozin IB, Kolchanov NA (1992) Somatic hypermutagenesis in immunoglobulin genes. II. Influence of neighbouring base sequences on mutagenesis. Biochim Biophys Acta 1171:11–18

Smith DS, Creadon G, Jena PK, Portanova JP, Kotzin BL, Wysocki LJ (1996) Di- and trinucleotide target preferences of somatic mutagenesis in normal and autoreactive B cells. J Immunol 156:2642–2652

Umar A, Schweitzer PA, Levy NS, Gearhart JD, Gearhart PJ (1991) Mutation in a reporter gene depends on proximity to and transcription of immunoglobulin variable transgenes. Proc Natl Acad Sci USA 88:4902–4906

Varade WS, Insel RA (1993) Isolation of germinal center-like events from human spleen RNA: somatic hypermutation of a clonally related V_H6DJ_H rearrangement expressed with IgM, IgG, and IgA. J Clin Invest 91:1838–1842

Varade WS, Marin E, Kittelberger AM, Insel RA (1993) Use of the most J_H-proximal human immunoglobulin heavy chain variable region gene, V_H6, in the expressed immune repertoire. J Immunol 150:4985–4995

Varade WS, Carnahan J, Kingsley P, Insel RA (1997) Inherent properties of somatic hypermutation as revealed by human nonproductive V_H6 immunoglobulin rearrangements (submitted for publication)
Wagner SD, Neuberger MS (1996) Somatic hypermutation of immunoglobulin genes. Annu Rev Immunol 14:441–457
Wilson M, Hsu E, Marcuz A, Courtet M, DuPasquier L, Steinberg C (1992) What limits affinity maturation of antibodies in Xenopus – the rate of somatic mutation or the ability to select mutants? EMBO J 11:4337–4347
Wu H, Kaartinen M (1996) Distribution and nucleotide biases of the somatic hypermutations in the functional k light chain gene of a human follicular lymphoma line. Scand J Immunol 43:193–201
Yélamos J, Klix N, Goyenechea B, Lozano F, Chui YL, Gonzalez Fernandez A, Pannell R, Neuberger MS, Milstein C (1995) Targeting of non-Ig sequences in place of the V segment by somatic hypermutation. Nature 376:225–228

Antibody Diversification in the Rabbit: Historical and Contemporary Perspectives

D.K. LANNING and K.L. KNIGHT

1 Introduction

Scientists have largely solved the century-old problem (KINDT and CAPRA 1984) of how the extraordinary range and specificity of the antibody response is generated. Multiple germline gene segments contribute to antibody diversity through combinatorial rearrangement, joining imprecision, and random insertion (N-diversity). Further diversity occurs through somatic hypermutation and/or somatic gene conversion of V(D)J genes, and by the pairing of immunoglobulin heavy and light chains. The comparative study of vertebrate immune systems shows that the number of germline V genes used in V(D)J gene rearrangement, as well as the mechanisms used in somatic diversification of V(D)J genes, varies among vertebrates. The rabbit has played an important role both in resolving the century-old problem of antibody diversity and in broadening the view of antibody diversification gained through mouse and human studies. In this chapter, we describe our current understanding of antibody diversification in the rabbit in the context of the historical contributions that rabbit studies have made to this field.

Department of Microbiology and Immunology, Stritch School of Medicine, Loyola University Chicago, Maywood, IL 60153, USA

2 Historical Perspective

Although several theories were advanced to explain the origins of antibody diversity, most immunologists advocated one of two diammetrically opposed views that differed in their estimates of the number of variable (V) genes contained in the germline. Advocates of germline theories envisioned a large number of V genes, perhaps even a separate germline gene for each antibody V region. Advocates of somatic diversification theories, on the other hand, contended that the number of germline V genes was very small, possibly as few as one, and that V region diversity was generated by somatic mutation. In germline theories, therefore, antibody diversity arose during phylogeny, whereas in somatic diversification theories, it arose during ontogeny.

2.1 V_H Allotypes and the Generation of Antibody Diversity

The discovery of rabbit allotypes by OUDIN in 1956 (1956a,b) had a major impact on the debate over the generation of antibody diversity. OUDIN identified the allotypes, denoted a1, a2, and a3, and these were later shown to be associated with the V_H region. Subsequently, rabbit allotypes proved to be valuable genetic markers, playing a central role in several important immunological discoveries, including allelic exclusion of immunoglobulin synthesis (PERNIS et al. 1965; WEILER 1965) and the demonstration that immunoglobulins were composed of two polypeptide chains encoded by unlinked genes (DUBISKI et al. 1962; STEMPKE 1964; FLEISCHMAN et al. 1962). The remarkable discovery of group **a** allotypes on several different immunoglobulin isotypes (TODD 1963; FEINSTEIN 1963; KINDT and TODD 1969; KOSHLAND et al. 1969) challenged a basic tenet of genetics, the one-gene–one-polypeptide chain theory. However, arguably the most important characteristic of rabbit group **a** allotypes, a characteristic that profoundly influenced the debate over antibody diversity, is their unique role as variable region markers. With the availability of genetic markers in the variable region of immunoglobulin heavy chains, investigators could directly address the central question of antibody diversity; that is, how many V regions are encoded in the genome. Most investigators considered group **a** allotypes to indicate the presence of a small number of germline V genes. Two lines of reasoning supported this view. First, the group **a** allotypes were inherited in simple Mendelian fashion. A total of 80%–90% of an individual rabbit's immunoglobulin heavy chains expressed group **a** allotypes, and these allotypes could be accurately predicted from the parental allotypes. Such allelic behavior is difficult to reconcile with a model postulating a large number of V genes, because it is then necessary to explain how the mutations conferring allotypy spread over a set of thousands of germline genes of differing specificities. Even if a plausible mechanism is postulated, a second difficulty immediately arises in explaining why meiotic recombination has not intermingled the different allotypic gene sets over time. Unless meiotic recombination were somehow suppressed, multiple crossovers between the members of allelic sets of V genes would rapidly render them indistinguishable. As a result of such considerations, most investigators regarded rabbit group **a** allotypes as support for a somatic mutation model of antibody diversification.

In view of these difficulties, many proponents of germline theories questioned whether rabbit group **a** allotypes were in fact V region markers. However, these objections were weakened when primary structural analyses showed a correlation between the allotypic specificities and distinct amino acid differences in N-terminal peptides, which contained the variable region, prepared from the heavy chain of rabbit immunoglobulin G (WILKINSON 1969; MOLE et al. 1971). MOLE et al. (1975) conclusively settled the issue by demonstrating the expression of group **a** allotypes on intact variable region fragments, Fv, obtained by papain digestion of rabbit heavy chain peptides. MAGE et al. (1984) later showed that the group **a** allotypic specificities correspond to amino acid differences in V_H framework regions 1 and 3.

2.2 Recombinant DNA Technology and Resolution of the Antibody Diversity Problem

In the late 1970s, researchers finally resolved the central genetic processes by which antibody diversity is generated by applying recombinant DNA technology. HOZUMI and TONEGAWA (1976) showed that functional immunoglobulin chain genes are produced by the joining of separate constant and variable region gene segments. Antibody diversification was achieved through contributions from a number of mechanisms, including combinatorial joining of a large number of V, (D), and J gene segments, junctional diversity at the interfaces between gene segments, somatic mutation, and combinatorial pairing of heavy and light chains. The model that emerged incorporated elements of both the germline and somatic diversification theories. There were, for example, multiple germline genes, although not nearly as many as envisioned by germline theories. Similarly, somatic mutation of variable region gene segments occurred, but it operated on a much larger number of germline genes than hypothesized by somatic diversification theories. The extraordinary diversity characterizing the antibody repertoire was generated both during phylogeny and during ontogeny. Thus, the data showed that the novel genetic mechanisms generating antibody diversity comprised both of the opposing viewpoints.

2.3 Inheritance of the V_H Allotypes

The difficulties that V_H allotypes had posed for germline theories, however, remained unanswered. Like the human and mouse genomes, the rabbit genome contained hundreds of V_H genes (GALLARDA et al. 1985; CURRIER et al. 1988). Yet, if most of these genes were used in VDJ gene rearrangements, how had the allotypic markers spread over the entire set of genes, and how was the integrity of these gene sets maintained against shuffling by meiotic recombination? Proponents of germline theories had advanced several explanations invoking such diverse mechanisms as gene expansion and contraction, crossover suppression, and inherited regulator genes (MAGE 1981; KINDT and CAPRA 1984). However, another explanation, which proved to be correct, was that only one, or a few, of the several hundred germline V_H genes was used in VDJ gene rearrangements.

2.4 Resolution of the V_H Allotype Problem: Preferential Utilization of V_H1

DIPIETRO and KNIGHT (1990) obtained evidence for the preferential utilization of a small subset of germline V_H genes by hybridization analysis of cDNA clones from splenic mRNA. They found that none of ten complete V_H-encoding cDNA clones hybridized to a V_Ha^- oligomer probe known to hybridize to more than 50% of approximately 100 cloned rabbit germline V_H genes, suggesting that V_H gene usage was not random. KNIGHT and BECKER (1990) subsequently demonstrated that the 3′-most V_H gene segment, *V_H1*, is preferentially utilized in rabbit VDJ gene rearrangements. They cloned and determined the nucleotide sequences of the 3′-most V_H genes from homozygous a1, a2, and a3 allotype rabbits. The translated V_H1 sequence for each allotype was identical to the sequence reported for the V_H framework regions (FRs) of pooled serum IgG of the corresponding allotype from normal, unimmunized rabbits (MOLE et al. 1971; MOLE 1975; JOHNSTONE and MOLE 1977; MAGE et al. 1984; KABAT et al. 1987). Furthermore, KNIGHT and BECKER showed through restriction mapping of cosmid clones that the V_H1 and V_H2 gene segments are missing in the mutant a2 rabbit, Alicia, because of a 10-kb chromosomal deletion. Early in life, Alicia rabbits, which are homozygous for the a2 heavy chain mutation *ali*, express almost no a2 immunoglobulin. Yet these rabbits express normal serum immunoglobulin levels because of a compensatory increase in the expression of V_Ha^- molecules (KELUS and WEISS 1986). These data, as well as evidence for preferential V_H1 usage in leukemic rabbit B cells (BECKER et al. 1990; KNIGHT and BECKER 1990), strongly suggested that allelic inheritance of the a1, a2, and a3 allotypes resulted from preferential utilization of allelic *V_H1* genes, *V_H1-a1*, *V_H1-a2*, and *V_H1-a3*.

3 Antibody Diversity in Rabbits: Somatic Gene Conversion and Somatic Mutation

In explaining the allelic inheritance pattern of V_Ha allotypes, the preferential expression of a single V_H gene segment posed a new problem: if the same V_H gene segment is used in 80%–90% of rabbit VDJ gene rearrangements, how is antibody diversity generated? BECKER and KNIGHT (1990) answered this question by comparing the nucleotide sequences of diversified *V_H1*-utilizing VDJ genes from homozygous a^3/a^3 rabbits with the nucleotide sequence of germline *V_H1-a3*. They identified clusters of nucleotide changes, including codon insertions and deletions, in the framework regions and in the complementarity-determining regions of the VDJ genes. Because such clustered nucleotide changes are characteristic of gene conversion events, they searched for, and in several instances identified, potential donors among the V_H gene segments 5′ of *V_H1-a3*. Two examples of gene conversion events in *V_H1* are shown in Fig. 1. These results strongly implicated a somatic gene conversion-like mechanism in the generation of antibody diversity in the rabbit.

Despite considerable investigation of gene conversion as a potential mechanism for generating antibody diversity in mammals, very little supporting evidence had been found previously (Krawinkle et al. 1983; Cumano and Rajewsky 1986). The observation of a gene conversion-like mechanism in rabbit was therefore unique among mammals (gene conversion-mediated diversification of bovine VDJ gene rearrangements has recently been reported (Parng et al. 1996)). However, diversification of rearranged V_H and V_L genes by somatic gene conversion had been described in the chicken (Reynaud et al. 1985, 1987, 1989; Thompson and Neiman 1987; McCormack and Thompson 1990). In the chicken, only the 3′-most V_H and V_L gene segments are functional, and all of the gene segments 5′ of these are pseudogenes. Antibody diversity is generated in chicken by gene conversion-mediated transfer of tracts of nucleotide sequence from V pseudogene donors into the rearranged functional VDJ gene. Gene conversion requires a high degree of sequence homology between the donor and recipient genes in the region upstream of the conversion event (McCormack and Thompson, 1990). The rabbit V_H gene segments that have been sequenced to date share high levels of sequence similarity and, unlike chicken V gene segments, about one half appear to be potentially functional (Becker et al. 1990). All of the cloned rabbit V_H gene segments are members of the V_HIII family (Currier et al. 1988). By definition, gene conversion is nonreciprocal homologous recombination. Although gene conversion events in the chicken leave the donor sequences unaltered (Carlson et al. 1990), the nonreciprocal nature of the mechanism employed by rabbits has not been demonstrated and is therefore referred to as gene conversion-like.

```
a       FR1
         G  L  F  K  P  T  D  T  L  T  L  T  C  T  V  S  G  F  S  L
VH1     GGTCTCTTCAAGCCAACGGATACCCTGACACTCACCTGCACAGTCTCTGGATTCTCCCTC
8.10    ..C..GA.......TGGA.GA.....................C...........A..A..
VH7     ..C..GA.......TGGA.GA.....................C...........A..A..

b       CDR1              FR2
         S  S  N  A  I  S  W  V  R  Q  A  P  G  N  G  L  E  W  I  G
VH1     AGTAGCAATGCAATAAGCTGGGTCCGCCAGGCTCCAGGGAACGGGCTGGAATGGATCGGA
4.7     ......T...G.G.G.....................................AC......
VH3     ......T...G.G.G.....................................AC......
```

Fig. 1a, b. Diversification of *V_H1* nucleotide sequence by gene conversion events in which V_H7 (**a**) and V_H3 (**b**) are potential gene donors for *V_H1*-utilizing VDJ genes 8.10 (a) and 4.7 (b), cloned from peripheral blood leukocytes of 8-week-old (**a**) and 4-week-old (**b**) rabbits

Somatic mutation has also been implicated in rabbit antibody diversification on the basis of extensive accumulation of point mutations in the D regions of VDJ genes from adult rabbits (DiPietro and Knight 1990; Short et al. 1991; Weinstein et al. 1994). Although the D regions of VDJ genes from newborn rabbits are identical to known D gene segments, the D regions of VDJ genes from adult rabbits bear no resemblance either to each other or to known D gene segments. It is unlikely that D regions are diversified by somatic gene conversion because, unlike the chicken, in which D region donor sequences are fused with upstream donor V pseudogenes (Reynaud et al. 1989), in the rabbit no potential D region donor sequences have been identified (Bernstein et al. 1985; McCormack et al. 1985; Currier et al. 1988; Fitts and Metzger 1990; Knight and Becker 1990; Roux et al. 1991; Raman et al. 1994). These data suggest that D regions are diversified by somatic mutation. However, an alternative mechanism, such as exonuclease activity followed by N addition, cannot be ruled out.

Strong evidence for somatic hypermutation at the rabbit heavy chain locus has been found downstream of rearranged VDJ genes (Lanning and Knight 1997). Point mutations 3′ of rearranged rabbit VDJ genes presumably can result only from somatic mutation because no potential gene conversion donors are known for this region. Lanning and Knight PCR-amplified the region immediately 3′ of rabbit VDJ genes from genomic splenic DNA and compared the nucleotide sequences of the cloned PCR products with the sequence of the germline J_H region. A high mutation frequency was found immediately 3′ of the VDJ genes, with a rapid decrease in the mutation frequency with increasing distance 3′ of the VDJ genes, as shown in Fig. 2. The distribution and types of mutations were similar to those that resulted from somatic hypermutation immediately 3′ of mouse VJ and VDJ genes (Weber et al. 1991, 1994; Lebecque and Gearhart 1990). These results strongly suggest that both somatic hypermutation and a somatic gene conversion-like mechanism contribute to the diversification of rearranged rabbit VDJ genes.

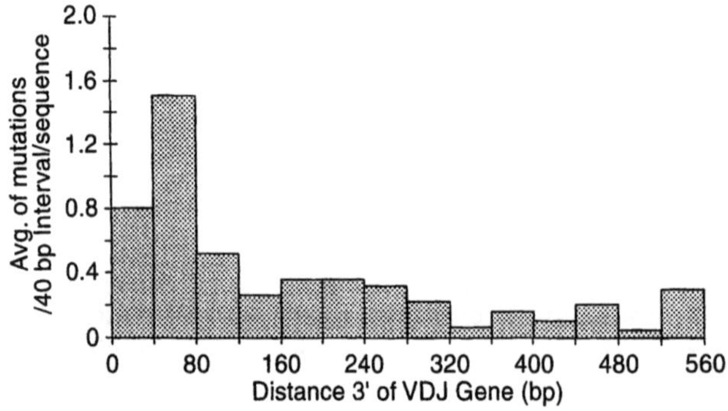

Fig. 2. Average number of mutations 3′ of VDJ genes cloned from peripheral blood leukocytes of 15-week-old rabbits. *n*, 21 sequences

4 Model for B Cell Development

By the early 1990s, there was enough evidence to indicate that the rabbit's unusual method of antibody diversification was just one aspect of an overall process of B cell development that was unique among mammals thus far studied. In 1994, KNIGHT and CRANE (1994) proposed a model for B cell development in the rabbit (Fig. 3). This model postulated that B lymphopoiesis is restricted to an early period during ontogeny and is absent or limited in adults. An individual rabbit's B cells are essentially all derived from an initial, self-renewing population generated early in life. Following VDJ gene rearrangement in the bone marrow, B cells migrate to gut-associated lymphoid tissue (GALT) at about 4 weeks of age, where they undergo proliferative expansion and diversify their VDJ genes through a somatic gene conversion-like mechanism and somatic hypermutation. On the basis of the anatomic site and developmental timing of B cell proliferation and VDJ gene diversification, the model postulates that these processes are driven by an exogenous factor(s), possibly a superantigen, derived from the microbial flora of the gut. The expansion of the B cell population and diversification of VDJ genes in GALT generates a primary antibody repertoire at about 6–8 weeks of age that serves for the lifetime of the rabbit. The major aspects of this model are now supported by a growing body of experimental evidence, as discussed in detail below.

4.1 Limited B Lymphopoiesis in Adult Rabbits

In experiments originally designed to determine V_H gene usage in $CD5^+$ B cells in rabbit, RAMAN and KNIGHT (1992) developed anti-CD5 mAb by cloning the rabbit CD5 gene and transfecting it into a murine T cell line. By immunofluorescence analysis of B lymphocytes from peripheral lymphoid tissues of adult rabbits they found that essentially all rabbit B lymphocytes were $CD5^+$. This observation strikingly contrasted with observations made in mouse and human, in which $CD5^+$ B lymphocytes constitute a small subpopulation of B cells. Since murine $CD5^+$ B lymphocytes develop exclusively during an early period in ontogeny and maintain themselves throughout an animal's lifetime by a process of self-renewal (HAYAKAWA

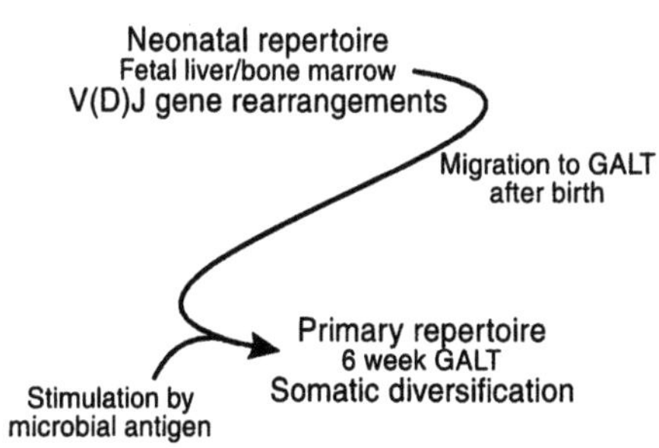

Fig. 3. Model of B cell development in rabbit. *GALT*, gut-associated lymphoid tissue

et al. 1985, 1986; Förster and Rajewsky 1987; Kantor et al. 1992), the expression of CD5 by essentially all rabbit B lymphocytes implied that rabbit B cells develop early in life and that B lymphopoiesis is absent or limited in adults.

While investigating the ontogenetic timing of VDJ gene diversification in rabbits, Crane et al. (1996) obtained additional evidence of limited B lymphopoiesis in adult rabbits. Although all rabbit VDJ genes are undiversified at birth (Friedman et al. 1994), they are extensively diversified by 7–8 weeks of age (Knight and Crane 1994). Crane et al. (1996) examined the extent of VDJ gene diversification at various timepoints in young rabbits by determining the nucleotide sequences of VDJ genes from peripheral blood leukocytes. The average number of nucleotide changes per V_H region increased progressively from three changes per V_H region at about 4 weeks of age to 12 changes by 6–8 weeks of age. Furthermore, although about 25% of the nucleotide sequences obtained from 4-week-old rabbits were undiversified, only one of 35 sequences (3%) from 6–8-week-old rabbits remained undiversified. These observations were extended by using an RNase protection assay to analyze a larger number of genes and to determine the level of undiversified V_H1 mRNA in appendix, peripheral blood, and bone marrow of both young and adult rabbits. Although a high level of undiversified V_H1 mRNA was found in young rabbits, only trace amounts of undiversified V_H1 mRNA were found in adults. The paucity of undiversified V_H1 mRNA in adult bone marrow was surprising since, if B lymphopoiesis were ongoing, one would expect to find undiversified VDJ genes from newly generated B cells. Crane and her colleagues therefore searched for evidence of B lymphopoiesis in adult bone marrow by PCR-amplifying signal joints from circular DNA excised during VD and DJ recombination. Although VD and DJ signal joints were readily PCR-amplified from the bone marrow of newborn rabbits, little or no PCR amplification of signal joints was obtained from adult bone marrow. These results indicate infrequent VD and DJ rearrangement in adult bone marrow and strongly suggest that B lymphopoiesis occurs to a limited extent in adult rabbits.

4.2 GALT as the Mammalian Bursal Equivalent

In the chicken, rearranged VJ and VDJ genes are diversified by somatic gene conversion after B cell migration to the bursa of Fabricius early in embryogenesis (Reynaud et al. 1987, 1989; Thompson and Neiman 1987; McCormack and Thompson 1990). The chicken bursa is structurally organized into numerous well-developed follicles, the microenvironment which supports B cell proliferation and the diversification process. Because this characteristic follicular structure is also found in rabbit GALT, particularly the appendix, it has been proposed that rabbit GALT may be the functional equivalent of the chicken bursa (Archer et al. 1963; Cooper et al. 1966; Knight and Crane 1994). Cooper et al. (1968) tested this idea by surgically removing GALT from newborn rabbits. These rabbits subsequently exhibited decreased serum Ig levels, fewer circulating lymphocytes, and diminished antibody responsiveness to several test antigens. Similarly, in a recent study, Vajdy et al. (1996) surgically removed the appendix, sacculus rotundus, and Peyer's patches from neonatal rabbits. The level of somatic diversification of IgM heavy chain VDJ genes was measured by determining the nucleotide sequences of IgM VDJ genes from peripheral B cells and comparing these nucleotide sequences to those of the appropriate germline V_H, D_H, and J_H gene segments. At 2–5 months of age, IgM VDJ genes in the GALT-less rabbits had undergone little diversification compared with those in control rabbits. Furthermore, the percentage of peripheral B cells was significantly reduced in the GALT-less rabbits compared with control rabbits. These studies of GALT-less rabbits support the view that rabbit GALT is a bursal homologue, functioning early in life as a site of B cell proliferation and VDJ gene diversification.

This view of rabbit GALT is further supported by the demonstration of ongoing diversification of rearranged VDJ genes by a gene conversion-like process that occurs in the rabbit appendix. WEINSTEIN et al. (1994) PCR-amplified rearranged heavy chain VDJ genes from B cells isolated from appendix germinal centers of 6-week-old rabbits. By cloning and determining the nucleotide sequences of VDJ genes from individual germinal centers, they found clonally related highly diversified sequences that could be traced back to a minimum of between one and four progenitors. These results strongly implicate appendix germinal centers as sites of rabbit B cell expansion and VDJ gene diversification. Furthermore, POSPISIL et al. (1995) reported evidence for preferential expansion and positive selection of B cells in the appendix on the basis of V_Ha allotypic framework region specificities. Compared with a2$^-$ B cells, a higher proportion of a2$^+$ B cells were found progressing through the cell cycle in the appendix of 6-week-old rabbits. In contrast, little cell proliferation and extensive apoptosis were observed in 6-week-old V_H-mutant *ali*/*ali* rabbits, in which the V_Ha allotype-encoding V_H1 gene is deleted. However, expanding populations of B cells expressing a2-like surface immunoglobulin were found in the appendix of 11-week-old *ali* mutants, suggesting the presence of upstream V_Ha2 allotype-encoding V_H genes that either have been rearranged or have contributed V_Ha2 FR specificities to rearranged a2$^-$ V_H genes through gene conversion. These data suggest that B lymphocytes are positively selected on the basis of FR1 and FR3 V_Ha specificities during their development in the appendix. Although the nature of the selecting agent is not known, suggestions have included the B lymphocyte cell surface glycoprotein CD5 (POSPISIL et al. 1996) and a component of the microbial flora of the gut (CRANE et al. 1996).

4.3 Role of Gut Microbial Flora in Rabbit B Cell Proliferation and VDJ Gene Diversification

The rabbit is unusual among animals known to generate their primary antibody repertoires through somatic diversification of a limited assortment of VDJ gene rearrangements. In rabbit, VDJ gene rearrangements are diversified postnatally rather than during fetal life. In the sheep and chicken, V(D)J gene rearrangements are diversified during fetal development in ileal Peyer's patches by somatic hypermutation or in the bursa by gene conversion, respectively (Reynaud et al. 1987, 1989, 1991, 1995; Thompson and Neiman 1987). Rabbit VDJ genes are diversified at 4–8 weeks of age by a somatic gene conversion-like mechanism and somatic hypermutation in GALT (Becker and Knight 1990; Weinstein et al. 1994; Crane et al. 1996). The timing and site of VDJ gene diversification in the rabbit suggests that initiation of the underlying molecular mechanism(s) depends upon exogenous antigen, perhaps derived from the microbial flora of the gut (Crane et al. 1996; Knight and Crane 1994). This view is supported by observations made with germfree rabbits. Štepánková and Kovárů (1978, 1985), for example, found markedly abnormal histologic development of the lymphoid organs in germfree rabbits. Although the thymus appeared to have developed normally in these animals, the appendix, sacculus rotundus, spleen, and mesenteric lymph node were poorly developed. The appendix lacked the well-developed follicular structure seen in normal appendix and contained far fewer lymphoblasts and lymphocytes. Furthermore, no germinal centers were apparent in the secondary lymphoid tissues before 3.5 months of age. In addition, Tlaskalová-Hogenová and Štepánková (1980) observed that germfree rabbits lacked natural antibacterial and hemolytic antibodies and were either unresponsive or deficient in their response to immunization with antigens. These observations suggest that normal microbial flora play an essential role in the development of rabbit lymphoid tissue and humoral immune responses.

Although a detailed picture of B cell development and antibody repertoire diversification in the rabbit is emerging, many questions remain unanswered. For example, what are the genetic mechanisms underlying restricted V_H gene usage and somatic gene conversion-mediated VDJ gene diversification? What is the molecular basis of B cell homing to GALT? What is the nature of the exogenous factor(s) driving B cell proliferation and VDJ gene diversification in GALT? These, and other unanswered questions, are exciting avenues of current research in rabbit immunology.

Vertebrates use a variety of approaches to generate their primary antibody repertoires. These approaches can be viewed as comprising two general strategies. In one strategy, used by human and mouse, an enormous number of antibody specificities are generated directly through V(D)J recombination because a large number of different V, (D), and J gene segments are rearranged. In a second general strategy, used by rabbit as well as chicken, sheep, and cow, a restricted range of antibody specificities is initially generated because of limited utilization of V, (D), and/or J gene segments, and these V(D)J genes are subsequently diversified through the targeted introduction of mutations into the rearranged V regions. Understanding the evolutionary and ecological significance of the strategies used by vertebrates to generate their primary antibody repertoires poses challenging and fascinating questions for the study of antibody diversification.

References

Archer OK, Sutherland DER, Good RA (1963) Appendix of the rabbit: a homologue of the bursa in the chicken? Nature 200:337–339
Becker RS, Knight KL (1990) Somatic diversification of immunoglobulin heavy chain VDJ genes: evidence for somatic gene conversion in rabbits. Cell 63:987–997
Becker RS, Suter M, Knight KL (1990) Restricted utilization of V_H and D_H genes in leukemic rabbit B cells. Eur J Immunol 20:397–402
Bernstein KE, Alexander CB, Mage RG (1985) Germline V_H genes in an a3 rabbit not typical of any one V_Ha allotype. J Immunol 134:3480–3488
Carlson LM, McCormack WT, Postema CE, Humphries EH, Thompson CB (1990) Templated insertions in the rearranged chicken IgL V gene segment arise by intrachromosomal gene conversion. Genes Dev 4:536–547
Cooper MD, Perey DY, McKneally MF, Gabrielsen AE, Sutherland DER, Good RA (1968) Production of an antibody deficiency syndrome in rabbits by neonatal removal of organized intestinal lymphoid tissues. Int Arch Allergy 33:65–88
Cooper MD, Perey DY, Gabrielsen AE, Sutherland DER, McKneally MF, Good RA (1966) A mammalian equivalent of the avian bursa of fabricius. The Lancet I:1388–1391
Crane MA, Kingzette M, Knight KL (1996) Evidence for limited B-lymphopoiesis in adult rabbits. J Exp Med 183:2119–2127
Cumano A, Rajewsky K (1986) Clonal recruitment and somatic mutation in the generation of immunological memory to the hapten NP. EMBO J 5:2459–2468
Currier SJ, Gallarda JL, Knight KL (1988) Partial molecular genetic map of the rabbit V_H chromosomal region. J Immunol 140:1651–1659
DiPietro LA, Knight KL (1990) Restricted utilization of germ-line V_H genes and diversity of D regions in rabbit splenic Ig mRNA. J Immunol 144:1969–1973
Dubiski S, Rapacz J, Dubiski A (1962) Heredity of rabbit gamma-globulin isoantigens. Acta Genet 12:136–155
Feinstein A (1963) Character and allotypy of an immune globulin in rabbit colostrum. Nature 199:1197–1199
Fitts MG, Metzger DW (1990) Identification of rabbit genomic Ig-V_H pseudogenes that could serve as donor sequences for latent allotype expression. J Immunol 145:2713–2717
Fleischman JB, Pain RH, Porter RR (1962) Reduction of the gamma-globulins. Arch Biochem Biophys [Suppl]1:174–180
Förster I, Rajewsky K (1987) Expansion and functional activity of Ly-1$^+$ B cells upon transfer of peritoneal cells into allotype-congenic, newborn mice. Eur J Immunol 17:521–528
Friedman ML, Tunyaplin C, Zhai SK, Knight KL (1994) Neonatal V_H, D and J_H gene usage in rabbit B-lineage cells. J Immunol 152:632–641
Gallarda JL, Gleason KS, Knight KL (1985) Organization of rabbit immunoglobulin genes. I. Structure and multiplicity of germ-line V_H genes. J Immunol 135:4222–4228
Hayakawa K, Hardy RR, Herzenberg LA, Herzenberg LA (1985) Progenitors of Ly-1 B cells are distinct from progenitors for other B cells. J Exp Med 161:1554–1568
Hayakawa K, Hardy RR, Stall AM, Herzenberg LA, Herzenberg LA (1986) Immunoglobulin-bearing B cells reconstitute and maintain the murine Ly-1 B cell lineage. Eur J Immunol 16:1313–1316
Hozumi N, Tonegawa S (1976) Evidence for somatic rearrangement of immunoglobulin genes coding for variable and constant regions. Proc Natl Acad Sci USA 73:3628–3632
Johnstone AP, Mole LE (1977) Sequence studies on the heavy chain of rabbit immunoglobulin A of differing a-locus allotype. Biochem J 167:255–267
Kabat EA, Wu TT, Reid-Miller M, Perry HM, Gottesman KS (1987) Sequences of proteins of immunologic interest. United States Department of Health and Human Services. Public Health Service, National Institutes of Health, Bethesda
Kantor AB, Stall AM, Adams S, Herzenberg LA, Herzenberg LA (1992) Differential development of progenitor activity for three B-cell lineages. Proc Natl Acad Sci USA 89:3320–3324
Kelus AS, Weiss S (1986) Mutation affecting the expression of immunoglobulin variable regions in the rabbit. Proc Natl Acad Sci USA 83:4883–4886
Kindt TJ, Capra JD (1984) The antibody enigma. Plenum, New York, pp 223–259

Kindt TJ, Todd CW (1969) Heavy and light chain allotypic markers on rabbit homocytotropic antibody. J Exp Med 130:859–866
Knight KL, Becker RS (1990) Molecular basis of the allelic inheritance of rabbit immunoglobulin V_H allotypes: implications for the generation of antibody diversity. Cell 60:963–970
Knight KL, Crane MA (1994) Generating the antibody repertoire in rabbit. Adv Immunol 56:179–218
Koshland ME, Davis JJ, Fujita NJ (1969) Evidence for multiple gene control of a single polypeptide chain: the heavy chain of rabbit immunoglobulin. Proc Natl Acad Sci USA 63:1274–1281
Krawinkle U, Zoelelein G, Bruggemann M, Radbruch A, Rajewsky K (1983) Recombination between antibody heavy chian variable-region genes: evidence for gene conversion. Proc Natl Acad Sci USA 81:4997–5001
Lanning DK, Knight KL (1997) Somatic hypermutation: mutations 3′ of rabbit VDJ H-chain genes. J Immunol, 159, (in press)
Lebecque SG, Gearhart PJ (1990) Boundaries of somatic mutation in rearranged immunoglobulin genes: 5′ boundary is near the promoter, and 3′ boundary is 1 kb from V(D)J gene. J Exp Med 172: 1717–1727
Mage R (1981) The phenotypic expression of rabbit immunoglobulins: a model of complex regulated gene expression and cellular differentiation. Contemp Top Mol Immunol 8:89–112
Mage RG, Bernstein KE, McCartney-Fransis N, Alexander CB, Young-Cooper GO, Padlan EA, Cohen GH (1984) The structural and genetic basis for expression of normal and latent V_Ha allotypes of the rabbit. Mol Immunol 21:1067–1081
McCormack WT, Laster SM, Marzluff WF, Roux KH (1985) Dynamic gene interactions in the evolution of rabbit V_H genes: a four codon duplication and block homologies provide evidence for intergenic exchange. Nucleic Acids Res 13:7041–7055
McCormack WT, Thompson CB (1990) Chicken IgL variable region gene conversions display pseudogene donor preference and 5′ and 3′ polarity. Genes Dev 4:548–558
Mole LE, Jackson SA, Porter RR, Wilkinson JM (1971) Allotypically related sequences in the Fd fragment of rabbit immunoglobulin heavy chains. Biochem J 124:301–318
Mole LE (1975) A genetic marker in the variable region of rabbit immunoglobulin heavy chain. Biochem J 151:351–359
Mole LE, Geier MD, Koshland ME (1975) The isolation and characterization of the V-H domain from rabbit heavy chains of different a locus allotype. J Immunol 114:1442–1448
Oudin J (1956a) Reaction de precipitation specifique entre des serums d'animaux de meme espece. C R Acad Sci 242:2489–2490
Oudin J (1956b) L'Allotypie de certains antigens proteidiques du serum. C R Acad Sci 242:2606–2608
Parng CL, Hansal S, Goldsby RA, Osborne BA (1996) Gene conversion contributes to Ig light chain diversity in cattle. J Immunol 157:5478–5486
Pernis B, Chiappino MB, Kelus AS, Gell PGH (1965) Cellular localization of immunoglobulins with different allotypic specificities in rabbit lymphoid tissues. J Exp Med 122:853–876
Pospisil R, Fitts MG, Mage RG (1996) CD5 is a potential selecting ligand for B cell surface immunoglobulin framework region sequences. J Exp Med 184:1279–1284
Pospisil R, Young-Cooper GO, Mage RG (1995) Preferential expansion and survival of B lymphocytes based on V_H framework 1 and framework 3 expression: "positive" selection in appendix of normal and V_H-mutant rabbits. Proc Natl Acad Sci USA 92:6961–6965
Raman C, Knight KL (1992) $CD5^+$ B cells predominate in peripheral tissues of rabbit. J Immunol 149:3858–3864
Raman C, Spieker-Polet H, Yam P, Knight KL (1994) Preferential V_H gene usage in rabbit Ig-secreting heterohybridomas. J Immunol 152:3935–3945
Reynaud C-A, Anquez V, Dahan A, Weill J-C (1985) A single rearrangement event generates most of the chicken immunoglobulin light chain diversity. Cell 40:283–291
Reynaud C-A, Anquez V, Grimal H, Weill J-C (1987) A hyperconversion mechanism generates the chicken light chain preimmune repertoire. Cell 48:379–388
Reynaud C-A, Dahan A, Anquez V, Weill J-C (1989) Somatic hyperconversion diversifies the single V_H gene of the chicken with a high incidence in the D region. Cell 59:171–183
Reynaud C-A, Garcia C, Hein WR, Weill J-C (1995) Hypermutation generating the sheep immunoglobulin repertoire is an antigen-independent process. Cell 80:115–125
Reynaud C-A, Mackay CR, Muller RG, Weill J-C (1991) Somatic generation of diversity in a mammalian primary lymphoid organ: the sheep ileal Peyer's patches. Cell 64:995–1005

Roux KH, Dhanarajan P, Gottschalk P, McCormack WT, Renshaw RW (1991) Latent a1 V_H germline genes in an a^2a^2 rabbit. Evidence for gene conversion at both the germline and somatic levels. J Immunol 146:2027–2036

Short JA, Sethupathi P, Zhai SK, Knight KL (1991) VDJ genes in V_Ha2 allotype-suppressed rabbits. Limited germline V_H gene usage and accumulation of somatic mutations in D regions. J Immunol 147:4014–4018

Stempke GW (1964) Allotypic specificities of A- and B-chains of rabbit gamma globulin. Science 145:403–405

Štepánková R, Kovárů F (1978) Development of lymphatic tissue in germfree and conventionally reared rabbits. In: Malek P, Bartos V, Weissleder H, Witte MH (eds) Lymphology. Thieme, Stuttgart, pp 290–294

Štepánková R, Kovárů F (1985) Immunoglobulin-producing cells in lymphatic tissues of germfree and conventional rabbits as detected by an immunofluorescence method. Folia Microbiol 30:291–294

Thompson CB, Neiman PE (1987) Somatic diversification of the chicken immunoglobulin light chain gene is limited to the rearranged variable gene segment. Cell 48:369–378

Tlaskalová-Hogenová H, Štepánková R (1980) Development of antibody formation in germ-free and conventionally reared rabbits: the role of intestinal lymphoid tissue in antibody formation to E. coli antigens. Folia Biologica 26:81–93

Todd CW (1963) Allotypy in rabbit 19S protein. Biochem Biophys Res Commun 11:170–175

Vajdy M, Sethupathi P, Knight KL (1996) Gut-associated lymphoid tissues and generation of the antibody repertoire. FASEB J 10:6 A182

Weber JS, Berry J, Manser T, Claflin JL (1991) Position of the rearranged V_κ and its 5′ flanking sequences determines the location of somatic mutations in the J_κ locus. J Immunol 146:3652–3655

Weber JS, Berry J, Manser T, Claflin JL (1994) Mutations in Ig V(D)J genes are distributed asymmetrically and independently of the position of V(D)J. J Immunol 153:3594–3602

Weiler E (1965) Differential activity of allelic γ-globulin genes in antibody-producing cells. Proc Natl Acad Sci USA 54:1765–1772

Weinstein PD, Anderson AO, Mage RG (1994) Rabbit IgH sequences in appendix germinal centers: V_H diversification by gene conversion-like and hypermutation mechanisms. Immunity 1:647–659

Wilkinson JM (1969) Variation in the N-terminal sequence of heavy chains of immunoglobulin G from rabbits of different allotype. Biochem J 112:173–185

Rabbit Appendix: A Site of Development and Selection of the B Cell Repertoire

R. Pospisil and R.G. Mage

1 Introduction

The normal rabbit primarily rearranges a single (V_H1) gene to several D_H and J_H genes (Becker et al. 1990; Knight and Becker 1990; Allegrucci et al. 1991). Thus combinatorial diversity is limited in developing rabbit B lymphocytes. Cohn and Langman (1990) hypothesized that in the mouse, a "high copy number repertoire" encoded by combinations of germline genes provides prompt and early protection against common pathogens. In the chicken, with only one rearranging V_H and V_L, the high copy number repertoire, is generated in the bursa of Fabricius by gene conversion (Langman and Cohn 1993). In view of the smaller contribution of combinatorial diversity to the B cell repertoire in rabbit compared to mouse, perhaps the rabbit also uses a gene conversion-like mechanism to diversify its rearranged V_H1-D-J gene sequences and generate its primary repertoire (Becker and Knight 1990; reviewed in Knight 1992). Work in our laboratory has shown that the appendix is a site of such diversification of rearranged VDJ sequences in young rabbits. Rearranged V_H sequences from individual appendix follicles reveal clonal diversification by gene conversion-like and somatic hypermutation mechanisms (Weinstein et al. 1994a). In addition to diversification of V_H gene sequences by gene conversion, positive and negative selection events occur in this organ that may involve endogenous as well as exogenous "superantigen-like" signals (Pospisil

Laboratory of Immunology, National Institute of Allergy and Infectious Diseases, Building 10, 11N311, 10 Center Drive, MSC 1892, National Institutes of Health, Bethesda, MD 20892–1892, USA et al. 1995, 1996a,b). These signals appear to lead to selective expansion of B cells bearing particular framework region sequences and apoptotic death of others. In the chicken, the bursa of Fabricius is the site within which B lymphocytes with rearranged V_H and V_L genes develop diversified V region sequences by a process of gene conversion that uses upstream pseudogenes as donors. Although the chicken bursa was originally thought to be a site of primary B cell development, we now know that B cells with already rearranged V genes migrate to the bursa, expand in number, and undergo primary repertoire development in this site (reviewed in Reynaud et al. 1994). Gene conversion alters the sequences of rearranged heavy and light chain genes. The finding of a role for the young rabbit appendix comparable to the role of the chicken bursa of Fabricius has renewed the view that the rabbit appendix is a mammalian bursal equivalent (Weinstein et al. 1994b).

2 Background

Figure 1 shows a working model of V_H repertoire development in the rabbit. The repertoire of fetal and neonatal rabbits is limited because of predominant utilization of only V_H1, although different D_H, J_H, and V_L are used. After some initial selection events in sites of B cell development such as fetal liver, bone marrow, or omentum, cells may migrate and populate the gut-associated lymphoid tissue (GALT) including, e.g., appendix, sacculus rotundus and Peyer's patches. There they expand in numbers and form germinal centers. These require gut antigens or superantigens as germfree rabbits do not develop normal germinal centers (Stepankova et al. 1980; Tlaskalova and Stepankova 1980). The B cell expansion and V_H gene diversification events that occur in the young rabbit appendix take place after birth and exposure to normal flora. This contrasts with the chicken bursa of Fabricius where B cell expansion and heavy and light chain variable region diversification initiates prior to hatching. Recent recombination activity is suggested by the finding of RAG-1 and RAG-2 expression in developing appendix cell subpopulations (Fuschiotti et al. 1997). In the 6-week-old rabbit appendix, gene conversion and mutation events are occurring that alter the sequences of rearranged V_H genes in cells that arrive with limited receptor diversity (Weinstein et al. 1994a). The sequences also develop point mutations. These changes extend into the D_H region. As there are no candidate donor genes for such changes, they are probably due to a non-templated hypermutation mechanism. We hypothesize that these early events in rabbit appendix represent the development of expanded B cell clones with potentially protective antibody specificities, the so-called high copy number repertoire (Cohn and Langman 1990). At this stage, conventional T lymphocytes with CD4 or CD8 are not detectable in appendix germinal centers. The primary repertoire that develops may be molded by antigens or superantigens, both endogenous and from gut flora. The repertoire may also be influenced by passively acquired maternal antibodies in serum, colostrum, and milk. Once the primary

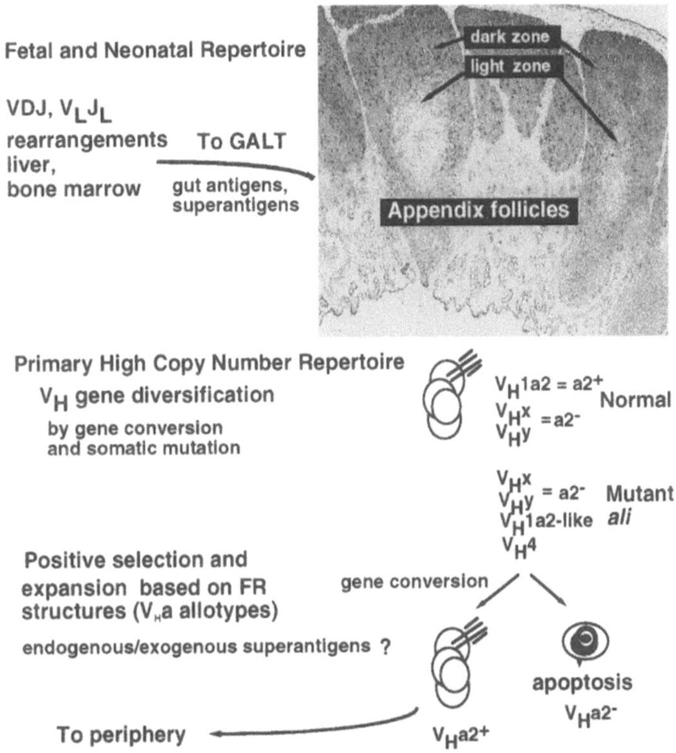

Fig. 1. A model of primary antibody repertoire development in the rabbit. The fetal and neonatal repertoire is limited because of limited V_H gene usage. Cells that develop in sites including fetal liver, bone marrow, or omentum may migrate to gut-associated lymphoid tissue (*GALT*; appendix, sacculus rotundus, Peyer's patches), where they expand in number. Germinal center (GC) development appears to be driven by endogenous and exogenous antigens or superantigens. By 6 weeks of age in normal rabbit appendix, B cells with rearranged V_H genes undergo clonal expansion and diversification by gene conversion-like and hypermutation mechanisms. In mutant *ali/ali* rabbits, appendix GC development is retarded. The appearance of cells with a2-like surface immunoglobulin appears to be due to gene conversion-like changes of other rearranged V_H genes. Cells with a2-like surface Ig preferentially expand in numbers and are less likely to undergo apoptotic death. Cells that survive selective events in appendix exit to the periphery where they may encounter foreign antigens

antibody repertoire develops in the GALT, it may be maintained by self-renewing $CD5^+$ B cells (Forster and Rajewsky 1987; Hayakawa et al. 1996), either in GALT or peripheral sites. The extent to which further diversification occurs during immune responses in peripheral secondary lymphoid tissues such as spleen, lymph nodes, and Peyer's patches is currently under investigation.

The *Alicia* mutation was discovered by Kelus and Weiss (1986). A small deletion at the 3′ end of the V_H gene cluster (Allegrucci et al. 1990) led to loss of V_H1 and one V_H pseudogene (Knight and Becker 1990). Although homozygous mutant *ali/ali* rabbits lack the V_H1a2 gene, gene conversion-like changes lead to B cells with a2-like surface Ig. We found that alterations in FR1 and FR3 sequences from splenic mRNA could be accounted for by gene conversion-like changes that utilized candidate donor sequences upstream of a rearranged V_H4 gene, the first functional gene in the mutants' V_H region cluster (CHEN et al. 1993, 1995). Our studies of the appearance of cells bearing a2-like epitopes in the appendix of mutant rabbits led us to the conclusion that there was positive selection and expansion based on framework region structures (V_Ha allotypes) (POSPISIL et al. 1995). We have suggested that the preferential expansion and survival of B cells based on FR1 and FR3 expression may involve "superantigen"-like interactions with endogenous and exogenous ligands. One endogenous ligand appears to be CD5 (POSPISIL et al. 1996a). A summary of these findings is presented in the following sections.

3 A Model for B Cell Selection in Appendix Germinal Centers

3.1 Superantigen-Like Unconventional Ag–Ab Interaction: "Positive" Selection of Rabbit Appendix B Cells

Figure 2 summarizes our findings on B cell selection in appendix germinal centers. B cells producing surface immunoglobulin with FR1 and FR3 V_Ha2 allotypic structures (reviewed in MAGE et al. 1984) are preferentially expanded and positively selected during their development in the rabbit appendix. We found that a higher proportion of V_Ha2^+ B cells were progressing through the cell cycle (S/G2/M) compared to $a2^-$ B cells, most of which were in the G1/G0 phase of the cell cycle (POSPISIL et al. 1995). The majority of appendix B cells in dark zones of germinal centers of normal 6-week-old rabbits stained with antibodies that detect proliferating cells [anti-Ki67 and proliferating-cell nuclear antigen (PCNA)] and there was very little apoptosis (Fig. 2). In contrast, in 6-week-old V_H-mutant *ali/ali* rabbits, there was little cell proliferation and significantly more extensive apoptosis. We found that even in the absence of V_H1, B cells with a2-like surface Ig developed and expanded so that the appendix of 11-week-old mutants resembled that of 6-week-old normal controls. By 11 weeks, the numbers and tissue localization of B cells undergoing apoptosis appeared similar to those found in 6-week-old normal appendix (POSPISIL et al. 1995). We hypothesized that B cells with a2-like molecules developed by gene conversion-like alterations of other rearranged V_H genes (CHEN et al. 1993, 1995). Recent studies of V_H sequençes expressed by FACS-sorted V_Ha2^+ appendix B cells from mutant rabbits support this proposal (SEHGAL et al., manuscript in preparation).

B cells with immunoglobulin receptors lacking the V_Ha2 allotypic structures appeared to be less likely to undergo clonal expansion and maturation. Although there are Bcl-2-independent mechanisms of regulating apoptosis as well as complex interactions between Bcl-2 family members, protection of cells against apoptotic death can be influenced by the level of Bcl-2 expression (reviewed in GAJEWSKI and THOMPSON 1996). We found that V_Ha2^+ B cells expressed high levels of Bcl-2

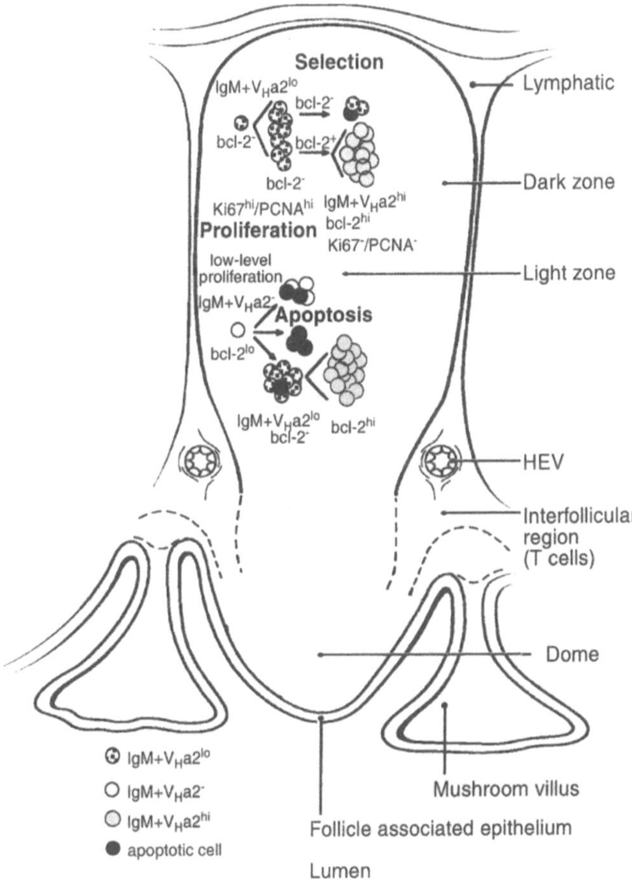

Fig. 2. Diagrammatic representation of a rabbit appendix follicle and B cell selection in germinal centers (GCs). The GC consists of a dark zone with large proliferating B cells stained with anti-Ki67 and proliferating-cell nuclear antigen antibodies and a light zone with small B lymphocytes. In 6-week-old normal rabbits V_Ha2^+ B cells predominate in the appendix. In contrast, < 10% of B cells in 6-week-old V_H mutants bear V_Ha allotypic specificities. In both normal and mutant rabbits, B cells with FR1 and FR3 V_Ha2 allotypic structures are preferentially expanded and positively selected in the appendix. V_Ha2^{hi} B cells also express high levels of Bcl-2 protein. In contrast, V_Ha2^- B cells have low level of Bcl-2 and more of them are undergoing programmed cell death. In mutant *ali/ali* rabbits, B cells with a2 molecules may develop by gene conversion-like alteration of other rearranged V_H genes. Those newly originated V_Ha2^{lo} B cells then preferentially proliferate but have no detectable Bcl-2. They are more likely to receive positive signals from endogenous or/and exogenous superantigens during the selection process in the appendix. If positively selected, they up-regulate a2 and re-express high levels of Bcl-2. *HEV*, high endothelial venule

protein compared to a2-negative B cells and suggested that B cells with FR allotypic motifs may become resistant to programmed cell death via the Bcl-2 pathway (Pospisil et al., to be published). We hypothesized that interactions of some as yet undefined foreign or endogenous antigens or superantigens with allotypic V_H FR1 and FR3 structures may provide signals for rabbit B cell proliferation and survival.

Comparisons of human and rabbit germline sequences suggested that amino acid positions associated with V_H allotypes in rabbit V_H FR1 and FR3 may contribute to unconventional superantigen-like binding interaction on rabbit B cells (Pospisil et al., to be published). In man, amino acids at positions 6–9, 74–78, and 85 are associated with the ability of staphylococcal protein A (SpA) to bind to V_H3 Fab fragments. Comparisons of human V_H3 and rabbit V_H sequences show highly conserved or identical residues at some positions. Thus, some residues in FR1 and FR3 of rabbit V_H may be associated with binding of superantigen(s). Surface IgM-positive B cells seeding the appendix probably receive signals to survive and proliferate from foreign antigens and superantigens. Exposure to exogenous superantigens may stimulate and expand most B cells that express surface Ig with rearrangements of the V_H1 gene. Upon contacting these antigens, B cells may be triggered to proliferate and this proliferative stimulus may be responsible for the formation of the lymphoid follicles in the appendix. In germ-free rabbits, the lymphoid compartment of the appendix does not develop, and the immune response is significantly reduced. Thus germinal center formation appears to be completely dependent on antigens from the gut (Stepankova et al. 1980; Tlaskalova and Stepankova 1980). The follicles then provide a microenvironment within which the rearranged VDJ genes of young rabbits diversify by gene conversion-like and hypermutation mechanisms (Weinstein et al. 1994a). Some of the B cells with highly diversified V_H gene sequences may exit directly to the periphery and contribute to the virgin B cell repertoire (see Fig. 1).

3.2 CD5 on Rabbit Appendix B Lymphocytes is a Potential Selecting Ligand for B Cell Surface Immunoglobulin Framework Region Sequences

Adult rabbits have a high proportion of cells with properties of self-renewal in the periphery and a low proportion of cells that are newly generated from bone marrow (Crane et al. 1996). This is in accord with the fact that allotype suppression is chronic in rabbits (reviewed in Mage 1974). If cells bearing allotypes are eliminated during the neonatal period, there is limited recovery of expression of the affected allotype during the life of the animal (discussed in Allegrucci et al. 1990; Mage 1993). Most dark zone B cells in appendix germinal centers express high levels of CD5 (Pospisil et al. 1996a) and essentially all peripheral B cells are $CD5^+$ (Raman and Knight 1992). The majority of B cells in normal animals bear V_Ha2 framework regions encoded by the V_H1 gene (Knight and Becker 1990; Allegrucci et al. 1991). Thus CD5 B cells express both CD5 and V_Ha2. The presence of both CD5 and V_Ha2 on the same cell raises the possibility of a relationship between the co-expression of these interacting proteins and the self-renewing capacity of these cells. We demonstrated an interaction between CD5 and V_H FR structures on B cells and postulated that CD5 is a potential selecting ligand that contributes to survival and expansion of B cells with V_Ha2^+ surface Ig (Pospisil et al. 1996a). The interaction between CD5 as a surface ligand and its receptor on the same or other

B cells may generate distinct activation signals at different stages of B cell development and selection. Figure 3 summarizes some of the possible roles of V_H–CD5 interactions.

It has been suggested that during human B cell development, superantigens or self-antigens interacting with evolutionarily conserved "family-specific" sequences in the FR1 and FR3 of the V_H, may significantly skew the composition of the B cell repertoire (KIRKHAM et al. 1992; SCHWARTZ and STOLLAR 1994; SILVERMAN 1994; ZOUALI 1995). Perhaps CD5–V_H interactions also play some role in selective expansion of human B cells expressing particular V_H structures.

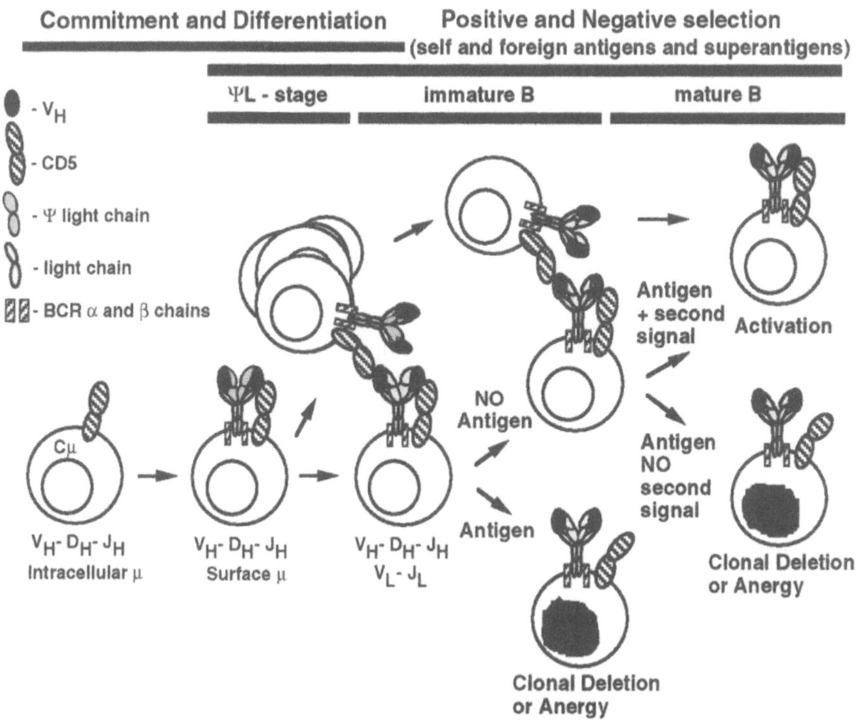

Fig. 3. Possible role(s) of CD5–V_H interactions. Since CD5 is expressed at early stages of B cell development it may play a role in the selection of B cells during the ΨL-chain stage and in maintenance and selective expansion of particular B cells during the immature stage when ΨL-chain is replaced with L-chain on the surface of the B cells. When B cells are passing through a tolerance-susceptible stage, those that recognize CD5, or other self-antigens with low affinity, could be stimulated but those recognizing foreign or self-antigens with high affinity would be eliminated. When mature B cells interact with antigen expressed on the surface of another cell, numerous cell–cell interactions are likely to occur. The relatively low affinity CD5–V_H interaction alone, or in combination with other signals on the same cell, or through interactions with nearby cells in a developing cluster, may affect the threshold for their activation. CD5–framework region interaction might also contribute to autostimulatory growth of transformed cells, as well as mediate selection of autoreactive repertoires

3.3 When and Where Does the Selection Take Place?

At birth the rabbit appendix has no organized follicular structures. The B cell repertoire appears to be very limited and newborn rabbits are relatively immunoincompetent. Rabbit appendix germinal centers (GCs) do not form until about 2 weeks after birth. Unlike the chicken bursa and the sheep ileal Peyer's patches where GC develop in the absence of antigen, the development of GC in the rabbit appendix appears to be antigen dependent. Until 4 weeks of age, the VDJ sequences found in the periphery remain relatively undiversified, but from then on the majority of V_H sequences appear highly diversified (CRANE et al. 1996). At 6 weeks after birth, rabbit appendix lymphoid tissue reaches its peak in both gross anatomical and GC follicle size and within individual germinal centers we found clonally related V_H sequences that were undergoing diversification (WEINSTEIN et al. 1994a,b). Recent studies indicate that self-renewal makes an important contribution to the B-lymphocyte compartment of adult rabbits as B lymphocytes of adult rabbits contain rearranged VDJ sequences that are highly diversified and there appears to be relatively little continuous B-lymphopoiesis in adult bone marrow compared to mouse (CRANE et al. 1996).

The GCs that arise in primary lymphoid tissue such as young rabbit appendix may be driven to develop by superantigen, self-antigen or other mediators of proliferation, but these reactions may not necessarily be specific for one particular antigen. Although GCs develop in the absence of antigen in the chicken bursa and sheep ileal Peyer's patches, foreign antigen is necessary in order for the GCs to reach their maximal sizes (REYNOLDS and MORRIS 1984; EKINO 1993). In germ-free rabbits, the lymphatic follicles grew very slowly and their enlargement was detectable only after 14 weeks (STEPANKOVA et al. 1980). In rabbits, there may also be an antigen-independent stage as in chicken, perhaps during the time when newborns are protected from foreign antigens by their mother's antibodies.

Perhaps during the first 2 weeks, in the presence of passive maternal antibodies, CD5 and other self-antigens that interact with FR1 and FR3 allotypic structural motifs can provide B cells with survival signals and limited expansion in the absence of stimulation by environmental antigens (Fig. 4). At 2–6 weeks of age, when the contributions from the mother's protective antibodies decline, foreign antigens and superantigens from gut flora stimulate B cell proliferation and formation of appendix germinal centers. Stimulation of B cells with both self and foreign antigens would then create follicles and provide a microenvironment within which VDJ genes of young rabbits diversify by gene conversion-like and hypermutation mechanisms. This could possibly lead to selection of some combining sites with high affinities for environmental antigens.

During B cell diversification, when new "anti-self" combining sites are created, the newly formed B cells bearing self-reactive Ig molecules are eliminated or develop clonal anergy. Prior to apoptotic B cell death, a second round of gene rearrangements may also occur to eliminate an "anti-self" combining site [receptor editing (GAY et al. 1993; TIEGS et al. 1993)]. Although B cells are positively selected during their development in the appendix (POSPISIL et al. 1996a), the amount of

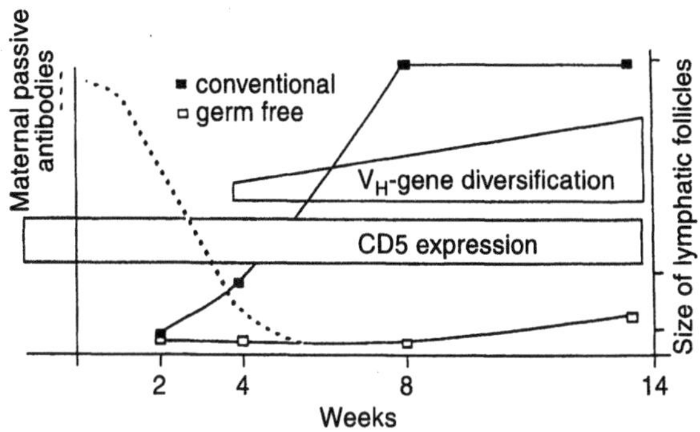

Fig. 4. The role of self and foreign antigens in the establishment of immunocompetence of newborn rabbits. In the absence of stimulation by foreign antigens, when newborns are protected by their mother's antibodies (or under germ-free conditions), CD5 and other self-antigens can provide B cells with stimulation and limited expansion. When maternal passive antibodies decline, antigens or superantigens from gut flora stimulate expansion of B cells and germinal centers develop. Newly created follicles then provide a microenvironment for V_H-gene diversification

signaling and qualitative differences in signaling may determine whether the B cells undergo negative or positive selection. Altering the signaling threshold through the exact same antigen receptor was shown to affect generation of B1 cells (Cyster and Goodnow 1995; Cyster et al. 1996). The stage of B cell development may determine the fate of a B cell upon antigen encounter due to intrinsic differences in B cell receptor signal transduction (Monroe 1996). As has been suggested for selection of T lymphocytes (Ashton-Rickardt and Tonegawa 1994), those B cells with relatively high affinity for a self-antigen may be eliminated but those that recognize self-antigens with low affinity may develop (Klinman 1996). We have shown that under conditions of competitive inhibition, V_Ha2-bearing $F(ab)'_2$ had a lower relative avidity compared to anti-CD5 (monoclonal anti-human CD5 antibody clone T1) for the site on CD5 recognized by this antibody (Pospisil et al. 1996a). Thus, the relatively low affinity CD5–V_H interaction alone may not induce apoptosis but rather a signal that is sufficient to promote expansion and/or survival of B cells and may influence the fate of B cell selection in combination with other signals. Once the functional antibody repertoire develops, it may be maintained by self-renewal of $CD5^+$ B cells. The site(s) of proliferation of such postulated self-renewing cells, their responses to foreign antigens, and the nature of further diversification that may occur in germinal centers of secondary lymphoid tissues are among the subjects currently under investigation.

4 Summary and Conclusions

As early as 1963, it was proposed that the rabbit appendix was a homologue of the chicken bursa of Fabricius (Archer et al. 1963). The finding that the young rabbit appendix was thymus independent contributed to the concept of central primary lymphoid tissue. Today we know that appendix is a site that generates the high copy number primary repertoire through diversification of rearranged V_H genes by gene conversion-like and somatic hypermutation mechanisms. Thus the appendix of young rabbits functions as a mammalian bursal equivalent. In the appendix, newly generated B cells also undergo selection processes involving self and foreign antigens and superantigens. Preferential expansion and survival of B cells in normal and mutant *ali* rabbits based on FR1 and FR3 expression may involve "superantigen"-like interactions with endogenous and exogenous ligands. One endogenous ligand appears to be CD5. Additional ligands may be produced by gut flora. Further studies in the rabbit model are needed to determine the fates of emigrants from primary GALT, their sites of postulated self-renewal in the periphery, and the nature of secondary diversification in secondary germinal centers where populations of B lymphocyte memory cells may develop. These data may also be helpful in understanding how the repertoire of human B cells is formed and how this repertoire might be manipulated for clinical benefit.

Acknowledgements. We thank Cornelius Alexander, Arthur O. Anderson, Hua Tang Chen, Joseph Dasso, Matthew Fitts, Patrizia Fuschiotti, Michael Mage, Enrico Schiaffella, Devinder Sehgal, Glendowlyn O. Young-Cooper, and Peter D. Weinstein for their contributions to the ideas and work presented in this review. We appreciate the editorial assistance of Shirley Starnes. R. Pospisil is on leave from the Institute of Microbiology, Academy of Sciences of the Czech Republic, Prague, Czech Republic.

References

Allegrucci M, Newman BA, Young-Cooper GO, Alexander CB, Meier D, Kelus AS, Mage RG (1990) Altered phenotypic expression of immunoglobulin heavy-chain variable-region (V_H) genes in Alicia rabbits probably reflects a small deletion in the V_H genes closest to the joining region. Proc Natl Acad Sci USA 87:5444–5448

Allegrucci M, Young-Cooper GO, Alexander CB, Newman BA, Mage RG (1991) Preferential rearrangement in normal rabbits of the 3′ V_Ha allotype gene that is deleted in Alicia mutants; somatic hypermutation/conversion may play a major role in generating the heterogeneity of rabbit heavy chain variable region sequences. Eur J Immunol 21:411–417

Archer OK, Sutherland DER, Good RA (1963) Appendix of the rabbit: a homologue of the bursa in the chicken? Nature 200:337–339

Ashton-Rickardt PG, Tonegawa S (1994) A differential-avidity model for T-cell selection. Immunol Today 15:362–366

Becker RS, Knight KL (1990) Somatic diversification of immunoglobulin heavy chain VDJ genes: evidence for somatic gene conversion in rabbit. Cell 63:987–997

Becker RS, Suter M, Knight KL (1990) Restricted utilization of V_H and C_H genes in leukemic rabbit B cells. Eur J Immunol 20:397–402

Chen HT, Alexander CB, Young-Cooper GO, Mage RG (1993) V_H gene expression and regulation in the mutant Alicia rabbit-rescue of V_Ha2 allotype expression. J Immunol 150:2783–2793

Chen HT, Alexander CB, Mage RG (1995) Characterization of a rabbit germline V_H gene that is a candidate donor for V_H gene conversion in mutant Alicia rabbits. J Immunol 154:6365–6371

Cohn M, Langman RE (1990) The protecton: the evolutionarily selected unit of humoral immunity. Immunol Rev 115:1–131

Crane MA, Kingzette M, Knight KL (1996) Evidence for limited B-lymphopoiesis in adult rabbits. J Exp Med 183:2119–2127

Cyster JG, Goodnow CC (1995) Protein tyrosine phosphatase 1C negatively regulates antigen receptor signaling in B lymphocytes and determines thresholds for negative selection. Immunity 2:13–24
Cyster JG, Healy JI, Kishihara K, Mak TW, Thomas ML, Goodnow CC (1996) Regulation of B-lymphocyte negative and positive selection by tyrosine phosphatase CD45. Nature 381:325–328
Ekino S (1993) Role of environmental antigens in B cell proliferation in the bursa of Fabricius at neonatal stage. Eur J Immunol 23:772–775
Forster I, Rajewsky K (1987) Expansion and functional activity of Ly-1$^+$ B cells upon transfer of peritoneal cells into allotype-congenic, newborn mice. Eur J Immunol 17:521–528
Fuschiotti P, Fitts MG, Pospisil R, Weinstein PD, Mage RG (1997) RAG-1 and RAG-2 in developing rabbit appendix subpopulations. J Immunol 158:55–64
Gajewski TF, Thompson CB (1996) Apoptosis meets signal transduction: elimination of a BAD influence. Cell 87:589–592
Gay D, Saunders T, Camper S, Weigert M (1993) Receptor editing: an approach by autoreactive B cells to escape tolerance. J Exp Med 177:999–1008
Hayakawa K, Hardy RR, Herzenberg LA (1986) Peritoneal Ly-1 B cells: genetic control, autoantibody production, increased lambda light chain expression. Eur J Immunol 16:450–456
Kelus AS, Weiss S (1986) Mutation affecting the expression of immunoglobulin variable regions in the rabbit. Proc Natl Acad Sci USA 83:883–886
Kirkham PM, Mortari F, Newton JA, Schroeder HW Jr (1992) Immunoglobulin V_H clan and family identity predicts variable domain structure and may influence antigen binding. EMBO J 11:603–609
Klinman NR (1996) The "clonal selection hypothesis" and current concepts of B cell tolerance. Immunity 5:189–195
Knight KL, Becker RS (1990) Molecular basis of the allelic inheritance of rabbit immunoglobulin V_H allotypes: implications for the generation of antibody diversity. Cell 60:963–970
Knight KL (1992) Restricted V_H gene usage and generation of antibody diversity in rabbit. Ann Rev Immunol 10:593–616
Langman RE, Cohn M (1993) A theory of the ontogeny of the chicken humoral immune system: the consequences of diversification by gene hyperconversion and its extension to rabbit. Res Immunol 144:422–446
Mage RG (1974) Altered quantitative expression of immunoglobulin allotypes in rabbits. In: Jerne N (ed) Current topics in microbiology and immunology. Springer, Berlin Heidelberg New York, pp 131–152
Mage RG (1993) Rabbit facts and diversification of V_H sequences by gene conversion. Comments on: A theory of the ontogeny of the chicken humoral immune system: the consequences of diversification by gene hyperconversion and its extension to rabbit. Ann Immunol 144:476–485
Mage RG, Bernstein KE, McCartney-Francis N, Alexander CB, Young-Cooper GO, Padlan EA, Cohen GH (1984) The structural and genetic basis for expression of normal and latent V_Ha allotypes of the rabbit. Mol Immunol 21:1067–1081
Monroe JG (1996) Tolerance sensitivity of immature-stage B cells. Can developmentally regulated B cell antigen receptor (BCR) signal transduction play a role? J Immunol 156:2657–2660
Pospisil R,Young-Cooper GO, Mage RG (1995) Preferential expansion and survival of B lymphocytes based on V_H framework 1 and framework 3 expression: "positive" selection in appendix of normal and V_H-mutant rabbits. Proc Natl Acad Sci USA 92:6961–6965
Pospisil R, Fitts MG, Mage RG (1996a) CD5 is a potential selecting ligand for B cell surface immunoglobulin framework region sequences. J Exp Med 184:1279–1284
Pospisil R, Young-Cooper GO, Mage RG (1996) Superantigen-like unconventional Ag–Ab interaction: "positive" selection of rabbit appendix B-cells. Vet Immunol Immunopathol 54:21–22
Raman C, Knight KL (1992) CD5$^+$ B cells predominate in peripheral tissues of rabbit. J Immunol 149:3858–3864
Reynaud CA, Bertocci B, Dahan A, Weill JC (1994) Formation of the chicken B-cell repertoire: ontogenesis, regulation of Ig gene rearrangement, and diversification by gene conversion. Adv Immunol 57:353–378
Reynolds JD, Morris B (1984) The effect of antigen on the development of Peyer's patches in sheep. Eur J Immunol 14:1–6
Schwartz RS, Stollar BD (1994) Heavy-chain directed B-cell maturation: continuous clonal selection beginning at the pre-B cell stage. Immunol Today 15:27–32
Silverman GJ (1994) Superantigens and the spectrum of unconventional B-cell antigens. The Immunologist 2:51–57
Stepánková R, Kováru F, Kruml J (1980) Lymphatic tissue of the intestinal tract of germ-free and conventional rabbits. Folia Microbiol 25:491–495

Tiegs SL, Russell DM, Nemazee D (1993) Receptor editing in self-reactive bone marrow B cells. J Exp Med 177:1009–1020

Tlaskalová-Hogenová H, Stepánková R (1980) Development of antibody formation in germ-free and conventionally reared rabbits: the role of intestinal lymphoid tissue in antibody formation to E. coli antigens. Folia Biol 26:81–93

Weinstein PD, Mage RG, Anderson AO (1994a) The appendix functions as a mammalian bursal equivalent in the developing rabbit. In: Heinen E (ed) Proceedings of the 11th International Conference on Lymphoid Tissues and Germinal Centers. Plenum, New York, pp 249–253

Weinstein PD, Anderson AO, Mage RG (1994b) Rabbit IgH sequences in appendix germinal centers: V_H diversification by gene conversion-like and hypermutation mechanisms. Immunity 1:647–659

Zouali M (1995) B-cell superantigens: implications for selection of the human antibody repertoire. Immunol Today 16:399–405

Affinity Maturation of the Primary Response by V Gene Diversification

D.M. TARLINTON[1], A. LIGHT[1], G.J.V. NOSSAL[1], and K.G.C SMITH[2]

1 Introduction

A key feature of the primary immune response to T cell-dependent antigens is the increase in the average affinity of antigen-specific antibody during the course of the response (EISEN and SISKIND 1964; SISKIND and BENACERAFF 1969). Indeed, the affinity of serum antibody for antigen in the late stages of the primary response is often equal to that of the secondary or memory response. The basis of high-affinity memory responses is well understood; V gene somatic hypermutation and B cell selection in the germinal center (GC) results in the generation of a population of recirculating memory B cells containing high affinity variants (GRAY 1994; RAJEWSKY 1996). The basis of affinity maturation of primary response serum antibody, however, is much less well understood. While it is often assumed that the process is very similar, if not identical, to that of the generation of memory B cells, few details have been determined. This chapter describes recent results from our laboratory and others which address the issue of affinity maturation of antibody-secreting cells in the primary response using the hapten (4-hydroxy-3-nitrophenyl)acetyl (NP) coupled to a protein carrier as a model system.

[1]The Walter and Eliza Hall Institute of Medical Research, Post Office, The Royal Melbourne Hospital, Victoria 3050, Australia
[2]University of Cambridge School of Clinical Medicine, Addenbrooke's Hospital, Hills Road, Cambridge CB2 ZQQ, United Kingdom

2 Affinity Maturation and Antibody Production in the Primary Response

Serum antibody production during the initial stages of a primary T cell-dependent response emanates from clusters of antibody-forming cells (AFCs) located outside the B cell follicles and adjacent to the T cell areas in secondary lymphoid organs (VAN ROOIJEN et al. 1986; JACOB et al. 1991). These foci of AFCs in the periarteriolar lymphoid sheath (PALS) initially secrete IgM, but subsequently switch to downstream isotypes (JACOB et al. 1991). The switch in Ig isotype probably occurs in a relatively synchronous fashion (NOSSAL and RIEDEL 1989). The antibody secreted by the AFCs of the foci is of low affinity (LALOR et al. 1992) and the V genes of these cells are rarely mutated (JACOB et al. 1993; MCHEYZER-WILLIAMS et al. 1993). After approximately 7 days, the frequency of antigen-specific AFCs in the spleen begins to decline (MCHEYZER-WILLIAMS et al. 1993; SMITH et al. 1996), which corresponds to the involution of the foci, as observed by histology (JACOB et al. 1991). Finally, during this period of AFC decline, the frequency of terminal deoxynucleotidyl transferase-mediated dUTP-biotin Nick End labeling (TUNEL)-positive $IgG1^+$ cells in the PALS increases (SMITH et al. 1996), strongly suggesting that the fate of the AFCs of the foci is apoptosis in situ. If this is the fate of the initial AFCs of the response, which cells secrete the high affinity antibody that eventually dominates the response? Where are they located? When and how are they generated?

The process of improving the affinity of antigen-specific B cells for the immunizing antigen has two components. One is selection for B cell clones utilizing the V_H-V_L combination(s) which has the highest intrinsic affinity for antigen. The second is the process of somatically mutating the V genes of antigen-specific B cells and then selecting for variants with improved affinity. This latter process usually occurs in the GC, although mutation and selection have been reported recently in mice unable to form GCs due to a deficiency in the cytokine tumor necrosis factor (TNF)-α (MARTIN et al. 1996). V gene somatic mutation and B cell selection have been analysed using several different approaches. Sequence analysis of the V genes of antigen-specific hybridomas made at various times after immunization has revealed a time-dependent increase in both mutation and the frequency of presumptive affinity-enhancing amino acid exchanges (GRIFFITHS et al. 1984; CUMANO and RAJEWSKY 1986). V genes recovered by PCR amplification from sub-populations of splenocytes, enriched for antigen-specific cells on the basis of cell phenotype, have defined the time of onset of mutation and that memory B cells have mutational loads equal to that of cells which participate in the secondary response (BEREK et al. 1991; WEISS and RAJEWSKY 1990). Direct recovery from histological sections of GC cells and AFCs from foci has demonstrated the localization of mutation to the GC (JACOB et al. 1990).

An alternative approach, pioneered in our laboratory, has been to purify antigen-specific B cells on the basis of their cell surface characteristics (MCHEYZER-WILLIAMS et al. 1991). This method has employed electronic cell sorting of isotype-switched B cells on the basis of their binding the immunizing antigen, in this case the hapten NP coupled to a fluorescent protein, allophycocyanin. In this way it has been possible to develop a phenotypic profile of the GC B cells, memory B cells and AFCs at specific times after immunization (LALOR et al. 1992; SMITH et al. 1996). Thus GC B cells can be uniquely identified on the basis of high level binding of the lectin peanut agglutinin (PNA) and high level expression of CD45R (B220). Similarly, AFCs are defined by expression of syndecan-1, recognized by the monoclonal 281.2 (LALOR et al. 1992) and by low level expression of CD45R (SMITH et al. 1996). An example of each of these phenotypes is shown in Fig. 1. This technology has enabled the analysis of antigen-specific cells of defined functional characteristics (i.e., GC-memory B cell and AFC) at various times after immunization, even more than 200 days after the primary challenge (SMITH et al. 1994). The ability to segregate GC-memory cells and AFCs has meant that the accumulation of somatic mutations in the V genes and the selection of these B cells on the basis of affinity in each of these compartments can be determined. This has allowed us to address the relationship between V gene somatic mutation in the GC, the continued generation of AFCs during the primary response, and affinity maturation of serum antibody.

3 Clonal Selection by Antigen

Preferential expansion of B cell clones expressing a V_H-V_L combination of intrinsically high affinity for the immunizing antigen is thought to be an important component in affinity maturation during a response. That is, given the enormous diversity of antigen specificity present in the naive B cell repertoire, Ig molecules possessing a range of affinities for any particular immunizing antigen presumably exist. In a situation of limiting antigen, as is the case with non-replicating antigens, those B cells which are activated and participate in the response presumably express Ig with the highest affinity for the antigen. As antigen levels decline during the course of the response, selection for B cells with the highest affinity for antigen presumably will become more intense. In the initial stages of the response the interclonal competition for antigen will be less intense than at later times since, at this time, antigen levels are maximal while serum levels of antigen-specific antibody and clonal expansion of antigen-specific B cells are minimal. However, there will still be competition since there is a finite amount of antigen. The consequences of this competition may well be reflected in the germline encoded composition of the Ig genes. This means that certain V_H and V_L segments may be selected, as may certain types of V(D)J junctions. This type of selection for the best germline-encoded specificity could therefore be reflected in the B cells preferentially expanded during the initial stages of the response, even prior to the onset of V gene somatic hypermutation and selection of variants in the GC. Evidence of such an outcome has already been presented in the elegant experiments of Kelsoe and colleagues (JACOB et al. 1993), who observed increasing restriction in the usage of V_H gene

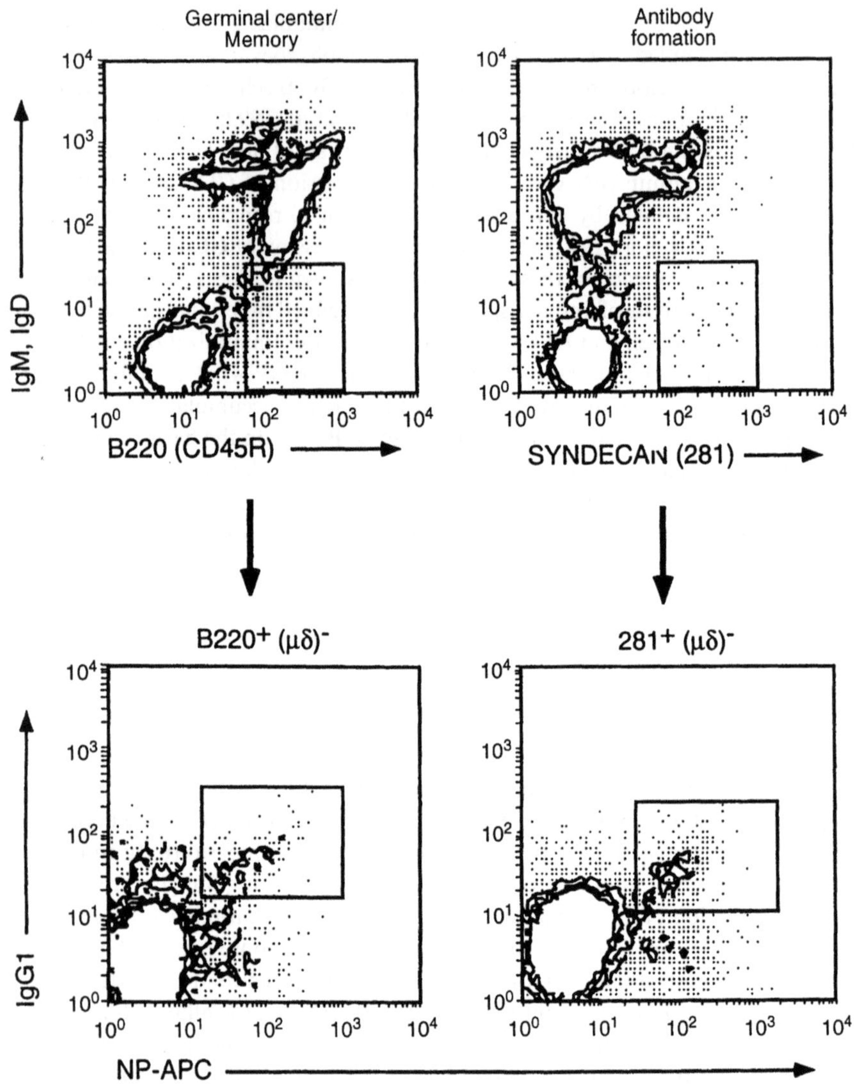

Fig. 1. Identification and isolation of antigen-specific B cells. Splenocytes from immunized mice were stained with the indicated antibodies coupled to various fluorochromes; anti-IgM and IgD, phycoerythrin; anti-B220 and anti-syndecan, fluorescein isothiocyanate; anti-IgG1, Texas Red; (4-hydroxy-3-nitrophenyl)acetyl, allophycocyanin. Gating on either $B220^{+}(IgMIgD)^{-}$ or $syndecan^{+}(IgMIgD)^{-}$ splenocytes reveals populations of isotype-switched germinal center B cells and antibody-secreting B cells, respectively

segments of GC B cells during the first week of the response to NP. Similarly, we reported that there appeared to be a restriction in the length of the CDR3 sequences between V_H genes of B cells in the GC compared to those in the foci (McHeyzer-Williams et al. 1993). Our results indicated that a preference for a nine-amino acid-long CDR3 was apparent by day 7 in GC B cells, while a greater range was permissible for AFCs at the same time point. One interpretation of this result was that recruitment into the GC was related to the nature of the CDR3, which in turn was related to the antigen binding properties of the antibody.

As part of a larger study of the molecular changes occurring in the AFCs of the primary response, we have had the opportunity to revisit this issue. Antigen-specific $IgG1^+$ B cells of both GC and AFC phenotype were sorted from spleen at various times after primary immunization. From these single, sorted B cells, cDNA was synthesized and the canonical V_H gene for the anti-NP response in IgH^b allotype mice was amplified by PCR. Analysis of the nucleotide sequence of such cells allowed us to examine selection of CDR3 lengths among GCs and AFCs during the early stages of the response. As shown in Fig. 2, in this analysis, there was no statistically significant difference in average CDR3 length at days 6, 10, or 14 between the two antigen-specific B cell populations. A trend, however, towards a more restricted CDR3 length in GC B cells compared to AFCs is apparent. That is, although the average for both populations is between approximately nine and ten amino acids at day 6, the shapes of the distributions are quite different. As time passes, CDR3-length distributions appear increasingly similar such that by day 14, they are essentially indistinguishable. The basis of this convergence will become apparent shortly. Thus, although we could not confirm our earlier finding as to the extent of the difference between AFCs and GC B cells in CDR3 length, our results are still suggestive of a more restricted distribution in GC B cells compared to AFCs. Perhaps the most important experiment is to measure the affinity of the antibody encoded by these two populations of B cells using a sensitive method. Since the original supposition was that any difference in the nature of these two populations was a consequence of differences in affinity for antigen (McHeyzer-Williams et al. 1993), this would seem the most accurate manner to resolve this point.

4 Somatic Mutation in Splenic Germinal Center B Cells

Sequence analysis of V_H genes of antigen-specific AFCs and GC B cells allowed an assessment of the contribution of somatic hypermutation to the diversification of the V gene repertoire of these two B cell populations. The accumulation of somatic mutation among GC B cells has been documented previously by ourselves and others. Our own earlier studies of the NP-specific response in C57BL-6 mice had shown that among GC B cells, mutation became apparent at around day 7 and then increased in frequency over the next 7 days, that being the duration of that particular study (McHeyzer-Williams et al. 1993). Concomitant with the increase in average V_H gene mutation frequency was an increase in the proportion of GC B cells containing a tryptophan-to-leucine exchange at position 33 of the V_H gene. This exchange has been shown by in vitro reconstruction experiments to increase

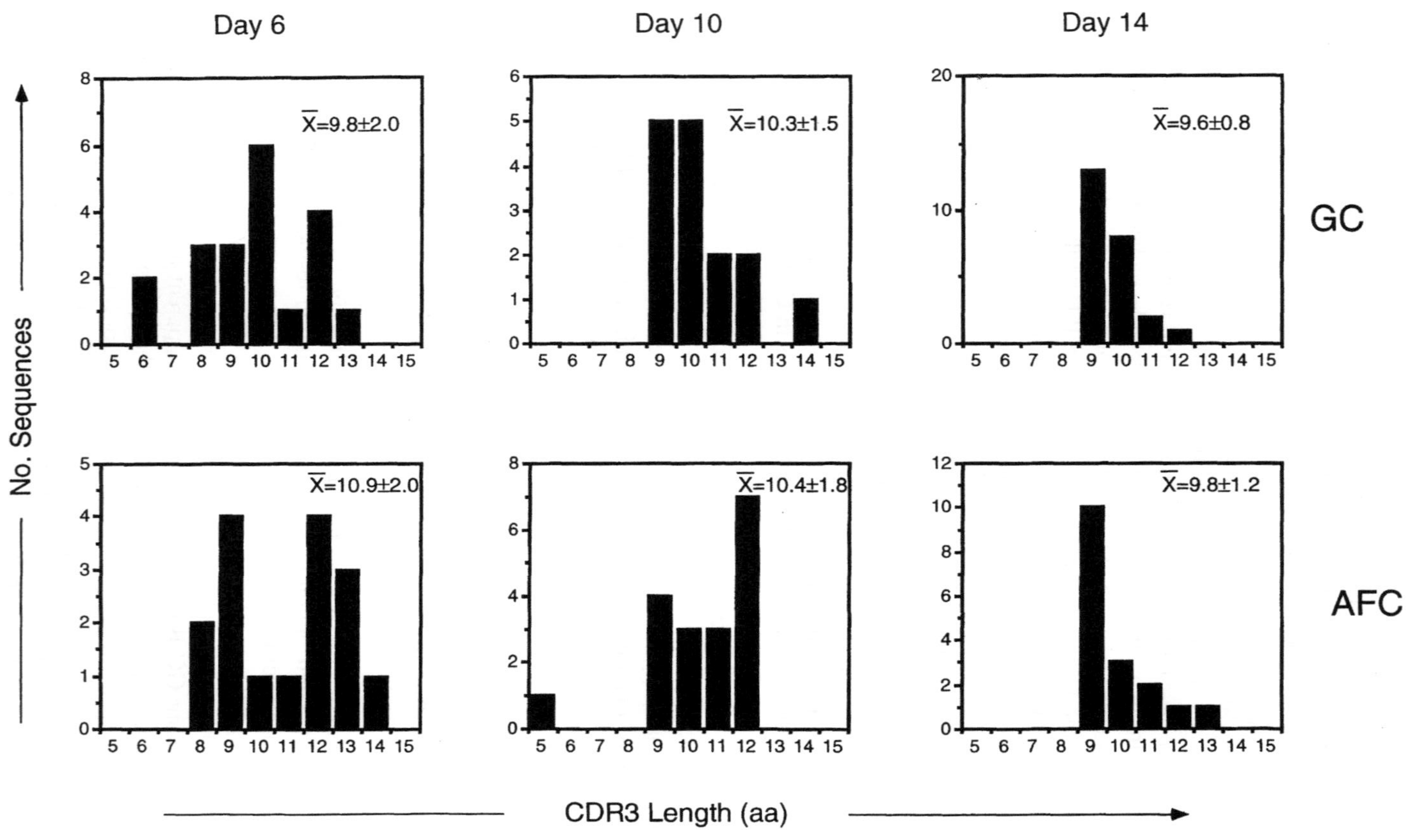

Fig. 2. Distribution of V_H gene CDR3 lengths in antigen-specific germinal center and antibody-secreting B cells early in the primary response. Nucleotide sequences were derived from single spleen cells isolated by flow cytometry from mice on the indicated number of days post immunization with (4-hydroxy-3-nitrophenyl)acetyl. Only V_H186.2 genes were considered. The number of sequences for each CDR3 length is plotted. The mean ± SD is given for each population. *GC*, germinal center; *AFC*, antibody forming cell

the affinity of antibody for NP by a factor of ten (ALLEN et al. 1988), to a level typical of the secondary response (CUMANO and RAJEWSKY 1986). That is, the frequency of this particular mutation can be used as a measure of the extent of B cell selection on the basis of affinity.

When the average frequency of mutations in the V_H genes of purified, antigen-specific GC B cells is measured as a function of time, it can be seen to initiate at around day 6, to increase in frequency during the next 14 days, and then essentially stabilize at a value equal to that of the long term memory population (Fig. 3), which is itself equivalent in mutation frequency to the AFCs generated during a secondary response (CUMANO and RAJEWSKY 1986; WEISS and RAJEWSKY 1990; RIDDERSTAD et al. 1996). Of some interest is the fact that the frequency of V_H gene mutation is initially ahead of V_H gene selection. Between day 6 and day 10 of the GC reaction in this particular response, the frequency of V_H gene mutation increases from an average of less than one nucleotide per gene to over four. The frequency of the tryptophan-to-leucine exchange at position 33, however, is 0% at day 6 and only 10% at day 10. Four days later, while the average mutation frequency has increased relatively little, the frequency of the position-33 exchange has increased five-fold, from 10% to about 50%. These data support the proposal of Kelsoe and colleagues (JACOB et al. 1993) who suggested that extensive mutation of V_H genes precedes efficient selection of high affinity variants, at least in this response. This may not be intrinsic to the mechanism of the GC, but rather reflect a change in some other physiological parameter, such as the titre of antigen-specific antibody and/or the amount of antigen. A decrease in the amount of antigen available may significantly alter the population dynamics of the GC, giving a considerable advantage to cells with higher affinity for antigen. The proportion of antigen-specific GC B cells containing V_H genes with the high affinity exchange at position 33 increases again between day 14 and day 17, indicative of continued selection of high affinity variants. After this time, however, both the frequency of V_H gene mutation and the proportion of genes containing the position-33 mutation does not increase to any substantial degree. This in turn suggests that affinity maturation of GC B cells is complete by around day 17, a value in agreement with that predicted by WEISS and colleagues (1992). In such a way a profile of affinity maturation of GC B cells has been developed.

5 Somatic Mutation in Splenic Antibody-Forming Cells

A critical question, however, is how affinity maturation in the GC relates to that of the antibody secreting cells of the primary response. To address this issue, we examined somatic mutation in splenic AFCs during the first 2 weeks of the primary response. NP-specific $IgG1^+$ $syndecan^+$ cells were sorted from spleen, cDNA synthesized from single cells, and $V_H186.2\text{-}C\gamma1$ rearrangements amplified by PCR. Positive PCR products were sequenced directly, without an intermediate cloning

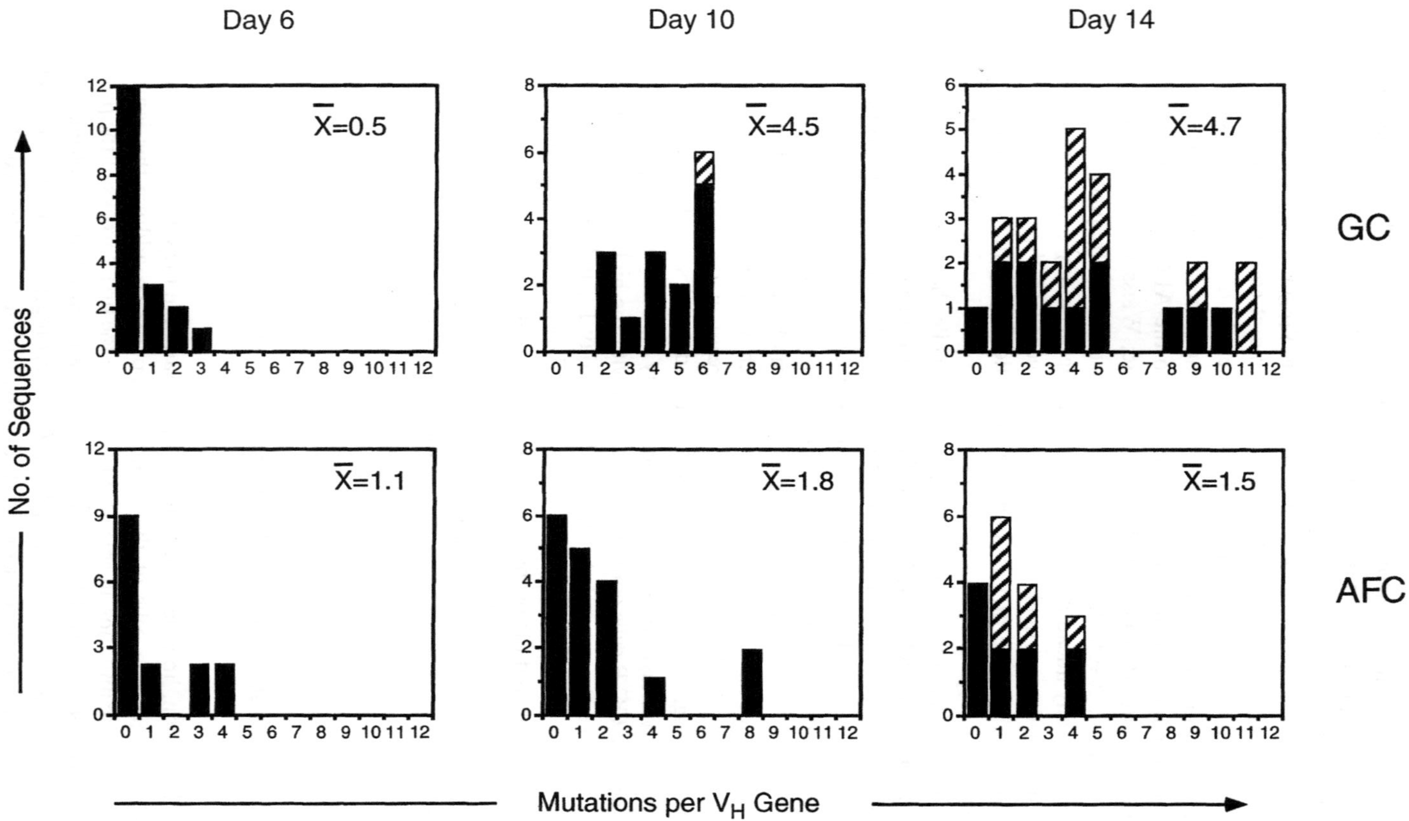

Fig. 3. Somatic mutation and selection in the primary response. Single cells of either an antibody-forming cell or germinal center phenotype were isolated on the indicated days and the nucleotide sequence of their V_H genes determined. The distribution of somatic mutations is plotted. Sequences containing the position-33 tryptophan-to-leucine exchange are indicated by the *hatched segment* of each column. *GC*, germinal center; *AFC*, antibody forming cell

step. This means that the background mutation frequency due to the methodology is essentially zero.

At 6 days following initiation of the immune response, the frequency of somatic mutations in the V_H genes of antigen-specific AFCs is similar to that in GC cells; an average of 1.1/gene in AFCs compared to 0.4 in GC B cells (see Fig. 3). Furthermore, both populations show no evidence of selection on the basis of improved affinity for antigen; none of the sequences in either population contain the position-33 tryptophan-to-leucine exchange associated with improved affinity for NP. The mutations appear to have a sporadic nature to them in that selection for amino acid replacement mutations in the CDRs and against such mutations in the framework (FW) regions is not apparent. Together these data would suggest that mutation starts at approximately day 6 and that it occurs in the B cells of both the GC and the foci. If mutation were to occur only in GC B cells it is difficult to explain how both populations of B cells have a similar distribution of mutations at such an early time point. In addition, the recent observation of somatic mutation in the apparent absence of GC formation in TNF-α knockout mice (MARTIN et al. 1996), would indicate that GCs are not obligatory for mutation.

At day 10 following immunization, a considerable difference is apparent in the frequency of V gene mutation in splenic GC and AFC B cells. Among antigen-specific AFCs, the frequency has remained at around one nucleotide per V_H gene, while in the GC B cells at the same time, the frequency has increased to be over four per gene on average (see Fig. 3). Selection on the basis of affinity is also apparent in the GC sequences in that one of the 15 sequenced cells shows the position-33 exchange. GC B cell sequences also show a preference for mutations encoding amino acid replacements in the CDR regions compared to the FW regions. No such preference is apparent in the distribution of mutations of the AFCs. Indeed, the distribution of mutations in the AFCs is very similar at day 10 to that at day 6. The fact that mutations do not accumulate in this B cell population between day 6 and day 10 has two possible explanations. First, mutation might only occur in AFCs during a narrow time period such that no additional mutations were introduced between day 6 and day 10. At face value, this would seem improbable. Second, the turnover of this B cell population might be such that cells do not persist long enough to accumulate a large number of mutations. Since persistence of mutated cells in the GC is thought to be associated with selection on the basis of antigen binding, this in turn would imply that such a selection does not occur for B cells of the foci. This would also fit with the distribution of the cell type thought responsible for B cell selection, the follicular dendritic cell (MACLENNAN 1994). The presence of $TUNEL^+IgG1^+$ cells in the PALS during this time is also consistent with rapid turnover in the AFC population (SMITH et al. 1996).

At day 14 following immunization, however, a change is apparent in the distribution of mutations in the V_H genes of antigen-specific AFCs in the spleen. While the average frequency is still low, 1.5 per V_H gene, a substantial fraction of sequences contain the tryptophan-to-leucine exchange at position 33. When the distribution of mutations in this B cell population is examined, it becomes apparent that the mutated sequences represent a highly selected population. First, of the 17 sequences, four retain germline V_H genes and may thus represent cells which have persisted from the extrafollicular foci. Of the mutated sequences 54% contain the position-33 tryptophan-to-leucine exchange, a proportion equal to that in the GC B cells at the corresponding time, despite the AFCs having a two-fold lower mutational load. That is, despite the lower frequency of somatic mutations among AFCs, there is an equal representation of the mutation which can, in isolation, confer high affinity binding of NP. While V_H gene sequences with similar mutation patterns exist in the GC, the key point is that the AFC population is enriched for such cells. The distribution of mutations in AFCs containing mutated V_H genes is a subset of the distribution in the GC, suggesting that such cells are selectively recruited from the GC into the AFC compartment. The basic point of the observed distribution of mutations in the V_H genes of day-14 splenic antigen-specific AFCs is the indication that B cells containing somatic mutations which confer high affinity are selectively induced to differentiate into AFCs. Our evidence for making this suggestion is the enrichment among AFCs of cells containing few somatic mutations in their V_H genes, but in which the single affinity-enhancing mutation at position 33 is very common. Not all day-14 splenic AFCs contain the affinity-enhancing mutation, so a proportion of these cells may represent AFCs of an extrafollicular origin.

6 Somatic Mutation in Bone Marrow Antibody-Forming Cells

The fact that the splenic AFCs could represent cells of two developmental origins, one being the extrafollicular foci and the other being the GC, complicates the interpretation of the results to some extent. For this and other reasons we have examined affinity maturation in the $IgG1^+$ antigen-specific AFCs of the bone marrow. The bone marrow (BM) has long been recognized as a site of long-term antibody production (Benner et al. 1981; Bachmann et al. 1994; Hyland et al. 1994), although in mice it has never been observed as a site of generation of an immune response. That is, structures associated with immune responses, such as GCs and foci of AFCs, have not been seen in mouse bone marrow. This suggests that AFCs which appear in the bone marrow after immunization do so as the result of emigration from another lymphoid organ. Indeed, in the secondary response to a model antigen, DiLosa and colleagues have observed the migration of AFCs from splenic GCs to the BM via the blood (DiLosa et al. 1991).

Examination of BM for the presence of antigen-specific AFCs using the enzyme-linked immunosorbent spot assay for cells secreting antibody in situ, revealed such cells from about day 7 onwards (Smith et al. 1996). The proportion of such NP-specific AFCs secreting antibody of high affinity was determined by measuring the frequency of AFCs which secreted antibody that could bind to a low haptenation plate coat. This revealed that the frequency of high affinity AFCs increased from 0% at day 8, to over 50% by day 14, and 100% by day 28 and beyond. Thus the BM AFC population of the primary response undergoes affinity maturation in much the same way as serum Ig. For this reason, we examined these AFCs for somatic mutation during the course of the primary response.

Application of the flow cytometric system previously used for the identification and isolation of NP-specific AFCs in the spleen to cells of the BM allowed for the recovery of $IgG1^+NP^+$ AFCs in an essentially pure form. RNA was extracted from these single antigen-specific BM AFCs at day 14 of the primary response, converted into cDNA and the V_H gene associated with the NP response amplified by PCR. Sequence analysis of such amplified genes revealed a striking distribution of somatic mutations among a subset of the V_H genes (Table 1). Of the 31 sequences examined, 25% were germline and presumably derive from cells secreting low affinity antibody. Among the remaining mutated sequences, the frequency of somatic mutations was very low for the time point, 2.5 per gene compared to an average for splenic GC cells of 4.5 at the same time. Amongst the mutated BM sequences, a remarkably high 96% contained the position-33 affinity-enhancing amino acid exchange, compared to approximately 50% in splenic GC B cells at the same time. The distribution of mutations in the BM AFC population is a more extreme example of the situation seen in the splenic AFCs at the same time point; few mutations but a high proportion of mutations targeted to produce an increase in affinity for antigen.

The low frequency of mutations in the BM AFCs at day 14 suggests that these cells are generated early in the GC reaction. It would also appear that once generated in the GC, the high affinity cells are induced to become AFCs which migrate to the BM. It is possible that the somatically mutated AFCs seen in the spleen at the same time represent the same population of GC emigrants, but which are either still in transit or are permanent residents of the spleen.

7 The Molecular Basis of Affinity Maturation in the Primary Response

The presence of a population of GC generated AFCs secreting high affinity antibody but containing a low frequency of mutations has a number of important

Table 1. Somatic mutation in day-14 antigen-specific B cells

		NP^+IgG1^+ B cells sorted from:		
		Spleen GC	Spleen AFC	BM AFC
Number of sequences		24	17	31
Sequences mutated (%)		96	76	74
Mutations per gene		4.7	1.5	1.9
Position. 33 tryptophan-to-leucine		54	41	71
R/S ratio	CDR1 and 2	14	>17	>38
	FWR 1–3	2.4	1.2	2.2

GC, germinal center; BM, bone marrow; AFC, antibody-forming cell; R/S, replacement to silent mutation ratio.

implications. First, it suggests that the AFCs responsible for production of high affinity antibody which drives affinity maturation of serum Ig in the primary response are generated throughout the course of the GC reaction and do not simply represent the differentiation of cells from the established memory B cell population. The distribution of mutations seen in the AFCs is not seen in the memory population (Weiss and Rajewsky 1990), or in the AFCs of the secondary response to NP (Cumano and Rajewsky 1986; Ridderstad et al. 1996). This in turn suggests that these high affinity AFCs are a distinct product of the GC reaction.

Affinity maturation of the antibody response has primarily been examined through the analysis of somatic mutation of GC B cells, despite the fact that these may not be the ones responsible for the secretion of antibody. Indeed, prior to the application of a flow cytometric technique capable of distinguishing AFCs and GC B cells on the basis of phenotype, it has been difficult to determine if there was a difference in the mutational pattern of V genes in AFCs compared to GC B cells. Our analysis of isolated AFCs during the primary response from both spleen and BM strongly suggests that a difference does exist in the manner in which they are generated, and this difference provides insights into how the GC functions to improve the affinity of antigen-specific B cells and generate a high affinity memory B cell population. Our results suggest that, during the early stages of the response, B cells within the GC which acquire high affinity as a result of particular somatic mutations are preferentially induced to differentiate into antibody secreting cells which leave the GC. Some of these cells migrate to the BM where they may reside for considerable times, secreting high affinity antibody. As the response progresses, we would propose that fewer of the high affinity variants are induced to differentiate into AFCs within the GC, but rather remain and eventually contribute to the memory B cell population. What remains to be determined is what controls the decision for a particular GC B cell as to whether it should differentiate into an AFC or remain in the GC for additional rounds of mutation and selection. One possibility is that this outcome represents one part of a control mechanism of the GC reaction involving levels of antigen, titres, and affinity of antigen-specific antibody and affinity of the mutated GC B cells. This is clearly an area for intensive research in the future.

Acknowledgements. The authors are grateful to Anna Ridderstad for critical review of this manuscript. Work reported in this article from the authors' laboratory has been supported by grants from the National Health and Medical Research Council (Canberra), by grant AI 03958 from the U.S. National Institute of Allergy and Infectious Diseases, and by a grant from the Human Frontier Science Program, Principal Investigator Diane Mathis.

References

Allen D, Simon T, Sablitzky F, Rajewsky K, Cumano A (1988) Antibody engineering for the analysis of affinity maturation of an anti-hapten response. EMBO J 7:1995–2001

Bachmann MF, Kündig TM, Odermatt B, Hengartner H, Zinkernagel RM (1994) Free recirculation of memory B cells versus antigen-dependent differentiation to antibody-forming cells. J Immunol 153:3386–3397

Benner R, Hijmans W, Haaijman JJ (1981) The bone marrow: the major source of serum immunoglobulins, but still a neglected site of antibody formation. Clin Exp Immunol 46:1–8

Berek C, Berger A, Apel M (1991) Maturation of the immune response in germinal centers. Cell 67:1121–1129

Cumano A, Rajewsky K (1986) Clonal recruitment and somatic mutation in the generation of immunological memory to the hapten NP. EMBO J 5:2459–2468

Dilosa RM, Maeda K, Masuda A, Szakal AK, Tew JG (1991) Germinal center B cells and antibody production in the bone marrow. J Immunol 146:4071–4077

Eisen HN, Siskind GN (1964) Variations in affinities of antibodies during the immune response. Biochemistry 3:996–1008

Gray D (1994) Regulation of immunological memory. Curr Opin Immunol 6:425–430

Griffiths GM, Berek C, Kaartinen M, Milstein C (1984) Somatic mutation and the maturation of the immune response to 2-phenyl-oxazalone. Nature 312:271–275

Hyland L, Sangster M, Sealy R, Coleclough C (1994) Respiratory virus infection of mice provokes a permanent humoral immune response. J Virol 68:6083–6086

Jacob J, Kassir R, Kelsoe G (1991) In situ studies of the primary immune response to (4-hydroxy-3-nitrophenyl)acetyl. I. The architecture and dynamics of responding cell populations. J Exp Med 173:1165–1175

Jacob J, Przylepa J, Miller C, Kelsoe G (1993) In situ studies of the primary immune response to (4-hydroxy-3-nitrophenyl)acetyl. III. The kinetics of V region mutation and selection in germinal center B cells. J Exp Med 178:1293–1307

Lalor PA, Nossal GJV, Sanderson RD, McHeyzer-Williams MG (1992) Functional and molecular characterisation of single (4-hydroxy-3-nitrophenyl)acetyl (NP)-specific, IgG1+ B cells from antibody secreting and memory B cell pathways in the C57BL/6 immune response to NP. Eur J Immunol 22:3001–3011

MacLennan ICM (1994) Germinal centers. Annu Rev Immunol 12:117–139

Matsumoto M, Lo SF, Carruthers CJL, Min JJ, Mariathasan S, Huang GM, Plas DR, Martin SM, Geha RS, Nahm MH, Chaplin DD (1996) Affinity maturation without germinal centers in lymphotoxin-alpha deficient mice. Nature 382:462–466

McHeyzer-Williams MG, Nossal GJV, Lalor PA (1991) Molecular characterisation of single memory B cells. Nature 350:501–505

McHeyzer-Williams MG, McLean MJ, Lalor PA, Nossal GJV (1993) Antigen-driven B cell differentiation in vivo. J Exp Med 178:295–307

Nossal GJV, Reidel C (1989) Sudden appearance of anti-protein IgG1-forming cell precursors early during primary immunization. Proc Natl Acad Sci USA 86:4679–4683

Rajewsky K (1996) Clonal selection and learning in the antibody system. Nature 381:751–758

Ridderstad A, Nossal GJV, Tarlinton DM (1996) The xid mutation diminishes memory B cell generation but does not affect somatic hypermutation and selection. J Immunol 157:3357–3365

Siskind GD, Benaceraff B (1969) Cell selection by antigen in the immune response. Adv Immunol 10:1–50

Smith KGC, Weiss U, Rajewsky K, Nossal GJV, Tarlinton DM (1994) Bcl-2 increases memory B cell recruitment but does not perturb selection in germinal centers. Immunity 1:803–813

Smith KGC, Hewitson TD, Nossal GJV, Tarlinton DM (1996) The phenotype and fate of antibody-forming cells of the splenic foci. Eur J Immunol 26:444–448

Van Rooijen N, Claasen E, Eikelenboom P (1986) Is there a single differentiation pathway for all antibody-forming cells in the spleen? Immunol Today 7:193–195

Weiss U, Rajewsky K (1990) The repertoire of somatic antibody mutants accumulating in the memory compartment after primary immunization is restricted through affinity maturation and mirrors that expressed in the secondary response. J Exp Med 172:1681–1689

Weiss U, Zoebelein R, Rajewsky K (1992) Accumulation of somatic mutants in the B cell compartment after primary immunization with a T cell-dependent antigen. Eur J Immunol 22:511–517

Lymphocyte Development and Selection in Germinal Centers

J. Przylepa[1], C. Himes[1], and G. Kelsoe[1]

1 Introduction

Germinal centers (GCs) represent an intense focal proliferation of antigen-specific B and T lymphocytes within a reticulum of follicular dendritic cells (FDCs). Newly formed GCs contain oligoclonal populations of lymphocytes that interact by cell-to-cell contact (Jacob et al. 1991a; Liu et al. 1991b; MacLennan 1994; Han et al. 1995a; Kelsoe 1995; Ferguson et al. 1996). These collaborations are necessary for the maintenance of the GC reaction, efficient activation of V(D)J hypermutation and selection, and the generation of memory B cells (Han et al. 1995a). By supporting these processes, the GC microenvironment controls and directs major pathways of antigen-dependent lymphocyte differentiation.

Recently, it has become clear that the mature, antigen-reactive lymphocytes in GCs express many properties of immature B and T cells. This cellular neoteny includes reactivation of the V(D)J recombinase, expression of components of the pre-B cell receptor, and sensitivity to apoptosis induced by autoreactivity (Han et al. 1995b; Pulendran et al. 1995; Shokat and Goodnow 1995; Han et al. 1996; Hikida et al. 1996; Zheng et al. 1996b). Together, these observations suggest that the antigen-driven diversification of V(D)J genes in GC lymphocytes recapitulates

[1]Department of Microbiology and Immunology, University of Maryland School of Medicine, 655 West Baltimore Street, Baltimore, MD 21201, USA

processes active in primary lymphoid tissues and may be homologous to that observed in gut-associated lymphoid follicles of those species in which immunoglobulin (Ig) genes undergo developmentally regulated hypermutation and/or gene conversion (ARCHER et al. 1963; THOMPSON and NEIMAN 1987; REYNAUD et al. 1987, 1994, 1995; BECKER and KNIGHT 1990; WEINSTEIN et al. 1994; KNIGHT and CRANE 1995; FUSCHIOTTI et al. 1997). In some way, the GC reaction seems to combine properties of primary lymphogenesis with antigen-dependent proliferation, selection, and differentiation.

In this chapter, we shall focus attention on a few key events in the GC reaction, especially those that may link programmed and antigen-dependent diversification of lymphocyte antigen-receptor genes.

2 The Role of Germinal Centers in V(D)J Diversification

Evidence for ongoing V(D)J mutation and selection in GCs is largely based on the construction of clonal genealogies from single GCs (JACOB et al. 1991b, 1993). In these experiments, antigen-specific GC cells were microdissected from histologic sections and their Ig heavy-chain gene rearrangements amplified by a polymerase chain reaction (PCR). The PCR generally amplified characteristic, or canonical, sets of V-to-J rearrangements without bias for specific clonality, i.e., the oligonucleotide primers employed flanked the V-to-J junction. This method permits in situ molecular genetic analysis of clonality and somatic diversification in lymphocyte populations. Nascent GCs contain B cell populations that are unmutated and clonally diverse. As the immune response progresses, the numbers of unique V-to-J junctional sequences [encoding the third complimentarity determining region (CDR3)] per GC decline and the frequency of point mutations increases. By day 10 of the response, only one or two distinct VDJ rearrangements are usually amplified from each GC, implying rapid and decisive clonal selection (JACOB et al. 1993). However, these clonally related VDJ fragments differ from one another [and from their germline sequence(s)] by the incorporation of numerous point mutations. Collections of shared and unique mutations in VDJ gene fragments containing a single CDR3 sequence can be organized into extensive family trees (JACOB et al. 1991b, 1993), demonstrating active clonal evolution in situ. V(D)J mutation in other histologic sites, followed by cellular immigration into GCs (for selection?) could not account for the oligoclonality and intraclonal diversification present in GCs. Oligoclonality in GC lymphocyte populations was first suggested by the specific assortment of antigen-binding B lymphocytes into different GCs (JACOB et al. 1991a; LIU et al. 1991b; VAN ROOIJEN et al. 1996); this oligoclonality has important consequences for humoral immunity.

2.1 Are Germinal Centers Necessary?

While these data define the GC microenvironment as a site for V(D)J hypermutation, is the GC the only site for this process? A significant body of evidence now suggests that V(D)J mutation, albeit at lower rates, may take place outside GCs. For example, Ig mutation occurs at low levels after immunization in *Xenopus* but these amphibians do not produce GC-like structures (WILSON et al. 1992). Indeed, phylogenetic studies imply that hypermutation in immunocytes is an ancient adaptation that arose coincidentally with V(D)J rearrangement (HINDS-FREY et al. 1993; GREENBERG et al. 1995). However, the character of V(D)J mutation in lower vertebrates is distinct from that observed in mammals and birds and may be mediated by some other (less elaborate?) mechanism.

In contrast, authentic Ig hypermutation and antibody affinity maturation have been observed in LT-α deficient mice (LT-$\alpha^{-/-}$) immunized with multiple intraperitoneal injections of antigen (MATSUMOTO et al. 1996). The splenic lymphoid architecture in these knockout animals is poorly organized and appears unable to support a typical GC reaction; other secondary lymphoid tissues such as lymph nodes and Peyer's patches are rare or absent (DE TOGNI et al. 1994; BANKS et al. 1995). Thus, antigen-driven hypermutation may well occur outside the GC, even in mammals. GCs may serve only as a catalyst for V(D)J mutation and/or selection. Observations of high rates of V(D)J mutation in some murine and human B cell lines (GREEN et al. 1995; ZHU et al. 1996; DÉNEPOUX et al. 1997) is consistent with this possibility.

However, continuing experiments with LT-$\alpha^{-/-}$ mice have revealed that lymph node-like structures are present in a variable fraction, from a few percent to almost one-third of animals (BANKS et al. 1995), and that GCs may be induced there. The cause(s) of this phenotypic leakiness is unknown but the possibility of V(D)J mutation and selection taking place in cryptic lymph nodes containing GCs can not be ignored. BEREK (SCHRODER et al. 1996) has demonstrated the formation of GCs in non-lymphoid tissues when chronic inflammation is present. These GCs contain T and B lymphocytes and even FDCs, but are apparent only by histologic inspection. Thus, while informative, the conclusion of MATSUMOTO et al. (1996) on the independence of Ig hypermutation and the GC reaction may yet be premature.

2.2 Do T Cells Mutate?

GCs that support V(D)J hypermuation and generation of the memory B cell compartment require the presence of antigen-specific, $CD4^+$ T cells (MILLER et al. 1995; YANG et al. 1996; ZHENG et al. 1996a). GC T cells are emigrants from the T cell zones of secondary lymphoid tissues [in the spleen, the periarteriolar lymphoid sheaths (PALS)] that, like GC B cells, intensely proliferate within the FDC reticulum (ZHENG et al. 1996a, b). During responses to pigeon cytochrome *c*, T cell clones that express a particular αβ TCR (Vα11/Vβ3) are selectively expanded in B10.A mice (MCHEYZER-WILLIAMS and DAVIS 1995). In GCs selection for this canonical TCR is hastened by preferential apoptosis in T cells bearing alternative TCRs (Zheng et al. 1996b). This programmed cell death is mediated by a Fas(CD95)/Fas-ligand(CD96) independent pathway that is associated with expression of Nur77, the zinc-finger transcription factor linked to TCR-mediated apoptosis (Liu et al. 1994; Woronicz et al. 1994).

During studies on TCR selection in GC T cell populations, we observed that Vα11 rearrangements amplified from GCs, but not the adjacent PALS, contained far more mutations than could be explained by polymerase errors (Zheng et al. 1994; Kelsoe et al. 1995). These mutations were similar to those present in Ig V(D)J in that they were confined to the TCRα variable region, exhibited similar nucleotide substitution patterns, and were preferentially targeted to one strand of DNA.

It is unlikely that these mutations represented PCR artifacts since accumulated mutations did not follow a Poisson distribution, identically mutated Vα11 rearrangements were independently amplified from adjacent tissue sections, and no mutated sequence could be explained by template chimerism. Furthermore, amplification of germline Vα11 exons from liver cells indicated that no unrecognized Vα11 element(s) was confused for mutated Vα11 genes. Similar mutations have been recovered from both T cell populations (approximately 20 cells) and single T lymphocytes dissected from GCs (Zheng et al. 1994, 1996b).

Our findings are controversial (Ford et al. 1994a; Bachl and Wabl 1995; McHeyzer-Williams and Davis 1995). These TCRα mutations were unusual not only by their mere presence, but also in their confinement to Vα rearrangements and over-representation in non-productive rearrangements.

The absence of V(D)J mutation in Vβ rearrangements is puzzling if TCRα genes are competent to mutate. If TCR mutation does represent a physiologic process, understanding its restriction to the a locus will require identification of the cis-acting elements that recruit the mutational machinery. However, it is remarkable how little DNA sequence appears to control hypermutation in κ transgenes (Wagner and Neuberger 1996).

The prevalence of Vα11 mutations in non-productive VαJα rearrangements can be explained by phenotypic selection (Zheng et al. 1994). In fact, GC T cells that express early markers of apoptosis frequently contain mutated Vα rearrangements; the great majority of these are in-frame (Zheng et al. 1996b). Thus if TCR mutation were most often deleterious, productively rearranged Vα mutants might be rapidly eliminated by negative selection. Primary GCs contain intensely dividing T cell populations that support high levels of apoptosis (Zheng et al. 1996b). However, this does not explain the high frequency of non-productive VαJα joins recovered in our studies. While we have no explanation for this, a minority of GC B cells reactivate the V(D)J recombinase. Could these non-productive VαJα rearrangements arise by illegitimate recombination? Time will resolve these issues. However, should TCR mutation occur in GCs, the most significant issue will remain: Why?

3 Selection in Germinal Centers

Although GCs often arise from genetically identical sister lymphocytes, the patterns of accumulating mutations in each is distinct (Jacob and Kelsoe 1992), precluding significant B cell trafficking between GCs and convergent selection driven by circulating antibody (Jacob and Kelsoe 1992; Vora and Manser 1995; Kelsoe 1996). Indeed, while the affinities of transfectoma antibodies encoded by V(D)J rearrangements from the same GC are similar, different GCs in the same spleen often support B cell populations with receptor affinities that differ 100-fold (Shimoda et al., unpublished). Each GC represents an independent experiment in clonal evolution that achieves a local fitness optimum.

3.1 All Selection Is Local

The significance of local selection is immediately obvious to fans of organized sport. All sport leagues have many local champions but only a single best team. For example, consider that the Nordic ski team of Enid, Oklahoma, and the M.I.T. rodeo equipage may at once be the best among their local competitors but fare poorly in comparison to more distant groups, say in Norway or Wyoming. Likewise, while each GC contains mutated B cells that have survived many rounds of local inter- and intraclonal competition (Bothwell et al. 1981; Crews et al. 1981; Griffiths et al. 1984; Manser et al. 1984; Wysocki et al. 1986; Blier and Bothwell 1987, 1988; Berek and Milstein 1987; Manser et al. 1987; Malipiero et al. 1987; Shlomchik et al. 1987; Allen et al. 1988; Sharon et al. 1989; Berek et al. 1991; Jacob et al. 1991b, 1992, 1993; McHeyzer-Williams et al. 1991, 1993; Berek and Ziegner 1993; Pascual et al. 1994), antibody affinities expressed by the mutant B cells of two GCs in the same spleen may differ significantly.

At first glance, the strategy of many independent tries in antigen-driven clonal evolution may not appear sound for achieving high-affinity, protective humoral immunity. Certainly many GCs late in the primary response do not contain B cells typical of memory responses (Jacob et al. 1993); occasionally, we have recovered mutated V_H rearrangements known to encode very low affinity antibody (Dal Porto et al., to be published) from GCs as late as 16 days postimmunization. However, this absence of efficiency may be offset by the potential benefit of diverse, independent evolutionary trajectories.

Consider the example of certain probable mutations, e.g., replacement exchanges at mutational hotspots (Kolchanov et al. 1987; Rogozin and Kolchanov 1992; Betz et al. 1993a, b; Smith et al. 1996; Wagner et al. 1995; Wagner and Neuberger 1996), that confer increased affinity but coincidentally promote self-reactivity. With global selection, these likely – and therefore early – mutations would tend to become fixed in all GC, while longer mutational pathways for affinity maturation would be lost. Subsequent deletion of increasingly autoreactive mutants (Han et al. 1995b; Pulendran et al. 1995; Shokat and Goodnow 1995) could make these early, high-affinity mutations evolutionary dead-ends and preclude further significant maturation of the antibody response. In contrast, the evolutionary diversity allowed by local selection would permit the continuation of affinity maturation by alternative mutational routes. (Imagine, if you will, the potential of husky M.I.T. undergraduates really interested in the physics of bull riding.)

Similar arguments (affinity-enhancing exchanges that render the antibody more sensitive to destabilization by later replacement mutations, etc.) for the utility of diverse evolutionary pathways in GC B cell populations come easily to mind. We wonder if the repertoire shifts described for several anti-hapten responses (MANSER et al. 1984, 1987; BEREK and MILSTEIN 1987; RAJEWSKY et al. 1987; LINTON et al. 1989; BEREK and ZIEGNER 1993) might not result from the replacement of early, "fragile" mutants by more robust, late-arriving competitors (MANSER et al. 1984, 1987).

Oligoclonality in GCs is the consequence of inter- and intraclonal competition between lymphocytes for proliferation and survival signals. Genetic and phenotypic evidence suggests that in GCs and T and B lymphocytes are selected by competition for antigen, presumably that reservoir maintained on the surface of FDCs (TEW et al. 1990). Selection generally favors lymphocytes expressing higher affinity antigen receptor molecules (B cells) or receptors that typically dominate the response (B and T cells) (GRIFFITHS et al. 1984; MANSER et al. 1984, 1987; BLIER and BOTHWELL 1987, 1988; RAJEWSKY et al. 1987; SHLOMCHIK et al. 1987; BEREK et al. 1991; MCHEYZER-WILLIAMS et al. 1991, 1993; BEREK and ZIEGNER 1993; FULLER et al. 1993; JACOB et al. 1993; KELSOE and ZHENG 1993; ZHENG et al. 1994, 1996a, b; RAJEWSKY 1996). In the primary anti-NP responses of *Igh*^{b} mice, early GCs contain between four and ten distinct VDJ joint sequences, indicating that each GC is founded by a variety of activated B cell clones (JACOB et al. 1993). By day 10 postimmunization, clonal diversity in GCs is reduced such that only one to two unique VDJ sequences are usually recovered. This reduction in the genetic variance of GC B cell populations is correlated with the loss of B cells expressing non-canonical V_H gene segments and/or VDJ rearrangements that do not make use of the characteristic D elements and CDR3 residues (DFL16.1 and YYGS, respectively) (JACOB et al. 1993). Early selection in GC B cells has also been observed as the reduction of CDR3 lengths to sizes optimal for antigen-building (MCHEYZER-WILLIAMS et al. 1993).

3.2 Selection Mechanisms

Much of the purifying, interclonal selection in GCs takes place prior to the accumulation of significant levels of V(D)J mutations, uncoupling the processes of selection and hypermutation (JACOB et al. 1993; MILLER and KELSOE 1995; MILLER et al. 1995). However, the rapid loss of non-canonical GC B cell clones is coincident with their down-regulation of membrane IgD and the anti-apoptotic protein, Bcl-2. Significantly, neither GC B cells nor T cells express levels of intracellular Bcl-2 that can be detected by immunohistology; this is in sharp contrast to the heavy labeling found in all other splenic lymphocyte compartments (HAN et al. 1997).

Like B cells, early GC T cell populations are genetically diverse but become highly selected by day 16 of the response. This selection of T cells recapitulates virtually all of the features observed for maturing, anti-hapten B cell responses: increasing representation of a few β- and α-chain pairs and selection for restricted CDR3 lengths and sequences (ZHENG et al. 1996a, b). However, this antigen-driven selection may not be solely based on increasing receptor avidity, as recent studies indicate that high-affinity (slow off-rates) TCR molecules may actually impair T cell activation by antigen (VALITUTTI et al. 1995).

Selection in GC lymphocyte populations can also be inferred from the analysis of V(D)J mutations. In GCs, the frequency of Ig rearrangements containing crippling mutations, i.e., the introduction of termination codons or replacements at invariant amino acid residues, is initially high (one-third of sequences recovered from NP-specific GCs on day 8) but falls to less than 3% by day 14 of the response (JACOB et al. 1993). With time, mutations also become increasingly prevalent in the CDRs. In part, this focusing reflects antigen-driven clonal selection (SHLOMCHIK et al. 1987; WEISS and RAJEWSKY 1990; BEREK et al. 1991) but recent work by NEUBERGER and his colleagues demonstrates that Ig (but not TCR) CDR sequences contain a high density of sequence motifs that promote hypermutation (WAGNER et al. 1995; WAGNER and NEUBERGER 1996). BEREK et al. (1991) were the first to show that the frequency of high-affinity B cell mutants becomes enriched in GC populations; many other groups have subsequently confirmed this important observation.

Positive selection for antigen-reactive B- and T-cell clones in GCs is thought to reflect the continued mediation of proliferation and survival signals by BCR and TCR molecules. In the face of declining reserves of antigen, only those B cells expressing receptors sufficiently avid to recover the minute amounts of antigen held by FDCs may be capable of presenting enough processed antigen to activate GC T cells. While this requisite avidity may be quite low in the absence of global competition, the GC reaction, Ig hypermutation, and generation of B cell memory depend upon continuing B/T collaboration via the CD40:CD40L and CD86:CD28 pathways (HAN et al. 1995b). GC lymphocytes that can not sustain these antigen-dependent interactions probably die or differentiate into short-lived effector cells (LIU et al. 1989, 1991a; HAN et al. 1995b; SHOKAT and GOODNOW 1995).

If plausible mechanisms for positive selection of GC lymphocytes seem obvious, it is not so clear why the mutational diversification of V(D)J genes within GCs does not regularly give rise to self-reactive cells (DIAMOND and SCHARFF 1984; CASSON and MANSER 1995a, b). One mechanism for preventing the rise of autoreactivity is implicit in the low clonal diversity of GC T and B cell populations and their requirement for regular cognate interactions. Note that mutant B cells that acquire reactivity for some new antigen present in the GC will be unable to collaborate with the antigen-specific helper T cells present there. This simple process of specificity cross-checking should prevent the expansion of most autoreactive mutants and also curtail broadening of the GC response to unlinked epitopes present in the immunogen (KELSOE 1995).

This mechanism would, however, be circumvented by V(D)J mutations that ensured T/B collaboration. For example, consider a mutated TCR Vα rearrangement that introduces reactivity for self-Igκ peptide. GC T cells bearing this receptor would be capable of collaborating with many activated B cells and would even rescue those that had independently achieved (other) autoreactive specificities. Thus it is significant that GC T and B cells become highly sensitive to Fas-independent apoptosis mediated by antigen-receptor crosslinking.

Introduction of soluble antigen into immune mice has been used by several groups to study the fate of GC B cells that acquire specificity for antigens not present on the FDC membrane (HAN et al. 1995b; PULENDRAN et al. 1995; SHOKAT and GOODNOW 1995). Soluble antigen induces extensive and rapid B cell death within GCs but not within the extrafollicular foci of antibody-secreting cells. Apoptosis is greatest in the mIg^+ centrocyte population present in the GC light zone and is receptor specific; injection of hapten coupled to an irrelevant protein induces apoptosis in hapten-specific B lymphocytes but spares adjacent cells reactive with the immunizing carrier protein. This sensitivity is present in both *lpr* ($CD95^{-/-}$) and *gld* ($CD96^{-/-}$) mutant mice and therefore can not represent the activation-induced cell death (AICD) pathway(s) common in mature peripheral lymphocytes (NAGATA and GOLSTEIN 1995). Indeed, antigen-induced apoptosis in GCs appears most similar to receptor-driven death in immature/transitional B cells (LINTON et al. 1991; CARSETTI et al. 1995; MONROE 1996).

The very high expression of the Fas death trigger by GC B cells (HAN et al. 1997) suggests a role for AICD in the selection of high-affinity B cells, perhaps by the integration of BCR and Fas signals. However, careful study of the GC reaction in *lpr* mutant mice that express little or no Fas, fails to reveal significant deficits in both the expansion and selection phases of the response. Table 1 gives the average serum antibody titers and numbers of GCs per histologic section observed in a small group of C57BL/6.*lpr* mice immunized with NP conjugated to chicken

Table 1. Serum antibody titers and the germinal center response in C57BL/6. *lpr* mice[a]. Each value represents the arithmetic mean of 3-5 tissue sections

	NP-Specific serum antibody			
	Total $\lambda 1^+$	IgGl	IgM	GCs/section
Control (unimmunized)	1:100[b]	< 1:100	1:800	
Day 10 (n = 2)	1:72 408	1:102 400	1:6 400	44
Day 14 (n = 2)	1:25 600	1:25 600	1:3 200	35
Day 18 (n = 1)	1:102 400	1:102 400	1:12 800	35

[a]C57BL/b *lpr* female mice were injected with 50 mg of NP conjugated to chicken γ globulin as described (Jacob et al., 1991a). Serum antibody titers and splenic GCs were quantitated by standard procedures (Jacob et al., 1993).

[b]Each titer represents the geometric mean of 2–4 titrations of 1–2 serum samples.

γ-globulin (NP-CG). These animals produced robust titers of $\lambda1^+$, NP-specific antibody that was predominantly of the IgG1 isotype. Although the spleens of unimmunized *lpr* mice contained significantly higher numbers of PNA^+ GCs (7 vs 0–2; see JACOB et al. 1991a) the frequency of GCs per histologic section was comparable to that observed in wildtype C57BL/6 controls. About 82% of the induced GCs in C57BL/6.*lpr* mice contained $\lambda1^+$ cells and exhibited typical histologic features.

Sequence analysis of VDJ rearrangements recovered from splenic GCs and plasmacytic foci (JACOB et al. 1991a) present in immunized *lpr* mice (Table 2) suggests that antigen-driven selection is intact despite the virtual absence of Fas. Just as in normal mice (JACOB et al. 1993), dominance of GC B cells expressing the canonical V186.2 V_H gene segment is established by day 10 postimmunization. The high frequency of V186.2 rearrangements in GCs at days 10 and 14 of this response represents selection; the initial activation of B cells expressing non-canonical VDJs is apparent in the large fraction of analogue V_H segments recovered from extrafollicular plasmacytes (Table 2).

V(D)J hypermutation is active in the GCs of immunized *lpr* mice (see Table 2) with the average frequency of point mutations in the V_H segment approaching 1.1% by day 14. Again, this frequency is comparable to that observed in the C57BL/6 strain (JACOB et al. 1993). Among the V_H mutations present in these GC B cells was the trytophan (W) → leucine (L) replacement at codon 33 of the V186.2 gene that increases affinity for the NP hapten ten-fold (RAJEWSKY et al. 1987; ALLEN et al. 1988). This mutation becomes frequent by day 14 of the response (see Table 2), indicating that affinity-driven selection is intact. VDJ rearrangements containing the W→ L mutation were also recovered from plasmacytic foci at day 14 of the response. In contrast to most VDJ fragments recovered from plasmacytes, these arrangements were heavily mutated (0.7%–1.0%) and shared CDR3 sequences common to the selected GC populations. Thus, while these mutants could represent hypermutation outside of the GC microenvironment, we feel that it is more likely that mutated plasmacytes appearing late in the response are GC emigrants.

Programmed cell death can be induced in immature, cortical thymocytes by low doses of several activating agents, including antibodies specific for the TCR/CD3 complex, bacterial superantigens, and glucocorticoids. Remarkably, GC T cells are equally sensitive to these agents (ZHENG et al. 1996b) and initiate a death program that elicits high levels of Nur77, the zinc-finger transcription factor linked to TCR-mediated apoptosis in immature thymocytes (LIU et al. 1994; WORONICZ et al. 1994). This apoptosis is independent of CD95 and CD96 and confined to the GC microenvironment. The majority of TCR mutations present in productive Vα rearrangements are recovered from $Nur77^+$ GC T cells, suggesting that negative selection is either the cause or consequence of TCR mutations (ZHENG et al. 1994; ZHENG et al. 1996b). Like thymus, the GC microenvironment seems capable of purging unwanted T lymphocytes by inducing programmed cell death.

Table 2. Population genetics of primary focus and germinal center B cells of C57BL/6. *lpr* mice[a]

	Ig H-chain gene rearrangements						
	V_H		D			V_H Mutations	
	V186.2 (%)	Analogue[c] (%)	DFL16.1 (%)	Other[d] (%)	P:nP	Frequency (%bp)	33W → L (%)
Day 10[b]							
foci	67	33	72	28	25:1	0.067	0
GCs	97	3	40	60	8:1	0.350	8
Day 14[b]							
foci	64	37	32	68	> 27:1	0.190	7
GCs	84	16	88	12	7.3:1	1.058	34
Day 18[b]							
GCs	58	42	29	71	23:1	0.729	29

[a]C57BL/6-*lpr* female mice were immunized with 50 mg of NP conjugated the chicken γ-globulin as described (Jacob et al., 1991a).
[b]Focus and GC cells were recovered by microdissection and the VDJ rearrangements they contained were amplified by a specific PCR (Jacob et al., 1991b).
[c]Analogue V_H genes recovered included: C1H4, V23, 24.8, 165.1, CH10 and V102 (Gu et al., 1991).
[d]D segments recovered included: DSP2, DQ52, and unknown.

4 Neoteny in Germinal Center Lymphocytes

As B lymphocytes proliferate in primary GCs they assume a surface phenotype that is unusual for peripheral cells. Murine GC B cells express high levels of heat-stable antigen [HSA; (HAN et al. 1997; KIMOTO et al. 1997)], bind the lectin, peanut agglutinin [PNA; (ROSE et al. 1980)], down-regulate mIgD expression, and are labeled by the LIP-6 and GL-7 monoclonal antibodies [LIP-6^{+}, GL-7^{+}; (HAN et al. 1996, 1997; KELSOE 1996)]. Intracytoplasmic staining reveals additional similarities including the expression of the V(D)J recombinase and the presence of λ5 mRNA (HAN et al. 1996). Many characteristics of pre-B and immature B cells are shared by B cells in GCs (HAN et al. 1997).

Interestingly, GCs contain a population of small B220^{lo} cells that contain most of the *Rag* and λ5 message present in GC B cells (HAN et al., to be published). These cells do not express CD43 (KIMOTO et al. 1997), nor do they contain terminal deoxynucleotidyl transferase (Tdt) protein or mRNA (HAN and KELSOE, unpublished data). These findings suggest that GC B cells may be similar to the pre-B II cells present in bone marrow but do not differentiate (further?) to resemble the CD43^{+} compartment of pro- and pre-B I cells (HARDY et al. 1991).

Comparison of B220^{lo} GC B lymphocytes to pre-B cells implies a specific role for the upregulated RAG proteins: de novo VJ rearrangement in the L-chain loci. Even this possibility reflects developmental immaturity. Work by NEMAZEE (TIEGS et al. 1993) and WEIGERT (GAY et al. 1993; RADIC et al. 1993) demonstrates the replacement of functional L-chain gene rearrangements by the reinduction or prolongation of recombinase activity in immature B cells. This "receptor editing" may be a response to self-reactivity and serve to promote efficient production of B lymphocytes while maintaining strict prohibition against autoreactive receptor specificities (reviewed in RADIC and ZOUALI 1996).

Although the possibility of L-chain replacement in GC B cells is intriguing, it is not at all obvious how this process would benefit the GC reaction. In contrast to the role proposed for receptor editing in the bone marrow (i.e., loss of specificity) new L-chains expressed by GC B cells would need to maintain or reconstitute specificity for antigen held on the FDC. It seems unlikely that random L-chain replacement could often accomplish this. Furthermore, if injection of soluble antigen is a valid model for the fate of GC B cells that acquire self-specificity by mutation (HAN et al. 1995b; PULENDRAN et al. 1995; SHOKAT and GOODNOW 1995), the rapidity of apoptosis in autoreactive GC B cells (4–6 h) would likely preclude rescue by even those rare replacements that produced an appropriate antibody specificity. These constraints suggest that the RAG-1 and RAG-2 proteins in GC cells may represent an epiphenomenon of the reactivation of apoptotic pathways that ensure self-tolerance (LINTON et al. 1991; CARSETTI et al. 1995; MONROE 1996).

The key to understanding the role of the V(D)J recombinase in GCs lies in the analysis of the B220^{lo} population of GL-7^{+}/PNA^{+} cells. If *Rag* (gene locus) expression in GCs were inconsequential, H- and L-chain gene rearrangements expressed in the B220^{lo} and B220^{hi} cell compartments should be similar. If B220^{lo} cells represent autoreactive mutants undergoing receptor editing, they should carry productive, mutated H-chain rearrangements distinct from those that avidly bind antigen and with no evidence of mutational debilitation. A third possibility, that RAG^+ GC B cells represent mutants that have lost affinity for the immunogen, predicts that $B220^{lo}$ cells will contain debilitating mutations and/or V(D)J rearrangements that are adequate but not optimal for antigen-binding.

This final possibility remains unproven but is supported by our recent sequence analysis of V_H rearrangements present in $B220^{lo}$ and $B220^{hi}$, $GL\text{-}7^+$ cells recovered from C57BL/6 mice immunized with NP-CG (HAN et al. 1997). $B220^{lo}$ GC cells contained very high frequencies of non-canonical VDJ rearrangements or mutationally inactivated rearrangements of the V186.2 gene segment; $B220^{hi}$ $GL\text{-}7^+$cells contained only canonical V_H genes and almost half of these carried affinity-enhancing mutations. These findings open the possibility that BCR revision by secondary V(D)J recombination is driven by distinct signals (reactivity or its loss) at different stages of B cell development.

5 Germinal Centers in Secondary Responses

It is widely assumed that memory B lymphocytes reenter GCs during secondary immune responses to undergo further evolutionary refinement by additional rounds of mutation and selection (BEREK and MILSTEIN 1987; LINTON et al. 1992; DECKER et al. 1995; KLINMAN 1996). Analyses of V_H and V_L in hybridoma lines established after primary and secondary immunization with oxazalone (phO_x) conjugates provide support for this idea by demonstrating ever increasing mutation frequencies (BEREK and MILSTEIN 1987). In contrast, rechallenge of adoptively transferred memory B cells specific for NP did not produce additional mutations, leading SIEKEVITZ et al. (1987) to propose that B cells become refractory to (V(D)J hypermutation after entering the memory cell compartment (i.e., following Ig class switch).

One way to reconcile these seemingly contrary findings is to propose that the great majority of V(D)J mutations introduced during the secondary response diminish the affinity of already selected memory B cell mutants. In other words, as the affinity of a particular H + L chain pairing increases during somatic evolution, the probability that additional mutations will enhance affinity declines. This rule of diminishing returns is quite general (ALLEN et al. 1988; CASON and MANSER 1995a, b; CHEN et al. 1992, 1995); in Ig hypermutation it is reinforced by codon usage that promotes replacement mutations in the CDRs of V_H and V_L gene segments (WAGNER and NEUBERGER 1996).

Interestingly, in response to both phO_x and NP, novel antigen-reactive clones appear after rechallenge. In secondary responses to phO_x, B cells expressing mutated V_H Ox-1 and V_κ Ox-1 rearrangements are largely replaced by clones expressing unrelated Ig genes (BEREK and MILSTEIN 1987). Likewise, secondary and tertiary responses to NP are characterized by increasing frequencies of κ L-chain antibodies unrelated to the initially dominant, $\lambda 1^+$ response (RAJEWSKY et al. 1987). These observations imply de novo recruitment or the selective expansion of small B cell populations that escape detection in the primary response; recruitment and/or expansion may be possible only when earlier rounds of mutation and selection have slowed the rate of affinity maturation in dominant clone(s).

Any model for clonal participation in the secondary GC reaction can be tested by simple observation. Thus, we have studied the somatic genetics of secondary GC B cell populations by microdissection of individual GCs and the amplification of the recovered VDJ rearrangements by specific PCR (JACOB et al. 1993). Briefly, C57BL/6 mice were immunized by an intraperitoneal injection of 50 mg of NP-CG in alum. Some of these mice were killed for analysis of primary GCs; the remainder were rested for 60 days and then received a second intraperitoneal injection of NP conjugate. Secondary GCs were studied to determine if they were reformed by mutated memory cell precursors or unmutated B lymphocytes newly recruited into the secondary response. Nascent GCs of the primary response are distinctly polyclonal: as many as ten unique CDR3 joint sequences can be recovered from a single GC at day 6 postimmunization (Fig. 1; JACOB et al. 1993). Additionally, this diverse population is not dominated by B cell clones expressing the canonical V186.2 V_H gene segment. Instead, it is largely comprised of B lymphocytes bearing the related V_H segments, V23, C1H4, CH10, and 24.8 (GU et al. 1991; JACOB et al. 1993). This high genetic variance abruptly ends by days 8–10 of the response with the nearly complete dominance of VDJ rearrangements containing the V186.2 gene segment and reduction of clonal diversity to only one or two unique CDR3 joint sequences per GC (see Fig. 1). V(D)J hypermutation is initiated at days 6–7 postimmunization (JACOB et al. 1993; MCHEYZER-WILLIAMS et al. 1993) and the frequency of accumulated V_H mutations increases monotonically over the next 10 days such that the majority of late (days 18–21) primary GCs support $\lambda 1^+$ B cells carrying V_H genes with between five and eight point mutations.

In contrast to primary responses, GCs form more rapidly after rechallenge with NP; by the fourth day of the secondary response large, well-formed GCs are present in the splenic follicles. These GCs frequently contain $\lambda 1^+$ B cells but the proportion of $\lambda 1^-$ GCs (approximately 38%) may be somewhat greater than that seen in primary responses (JACOB et al. 1991a). Genetic analysis of secondary, $\lambda 1^+$ GC cells indicates that these populations are not clonally related to the mutated, canonical B cells that achieved numerical priority during the primary response (see Fig. 1). Indeed, while VDJ fragments containing the V186.2 segment are common early in the secondary GC response, they are quickly replaced by rearrangements of the analogue segments V23, C1H4, CH10, and 24.8. Coincidentally, the frequency of unique CDR3 sequences falls from three to four per GC at day 4 to one or two by day 14 postimmunization. These selected, non-canonical B cells are capable of V(D)J hypermutation, as about half of the V_H analogues acquire between one and two point mutations by day 6; the frequency of V_H mutations in this cell set increases steadily thereafter.

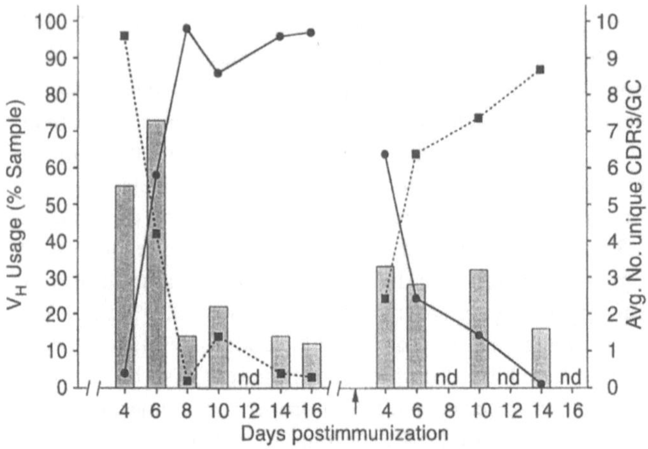

Fig. 1. Population genetics of germinal center B cells after primary and secondary immunizations. Female C57BL/6 mice ($n=28$) were injected intraperitoneally with 50 mg of NP coupled to chicken γ-globulin (Jacob et al. 1991a). At various times after the first and second (↑) immunizations, spleens from between two and four mice were taken for histology; cells were dissected from single $\lambda 1^+$ PNA^+ GCs and the recovered VDJ fragments amplified in a specific polymerase chain reaction (Jacob et al. 1991a, b). Cloned VDJ fragments ($n=30$–60 for each time point) were subjected to automated DNA sequencing to determine V_H gene segment usage and the average number of unique CDR3 joint sequences per GC (Jacob et al. 1993). V_H gene segments are grouped as V186.2 (■) and related analogue (Gu et al. 1991) genes (■) including, V23, CH10, C1H4, 24.8, V105, and 102.1. The average number of unique CDR3 sequences per GC (*shaded bars*) represents analysis of between two and five individual GCs. In the primary response, GCs initially contain diverse clonal populations that are not dominated by B cells expressing the canonical V186.2 V_H segment. However, by day 8 postimmunization, only between one and two unique CDR3 sequences can be recovered from most GCs, and VDJ rearrangements containing the V186.2 segment comprise >85% of amplified DNA. In contrast, secondary GCs seem to be established by fewer clones; on average, between three and four distinct CDR3 sequences are present in secondary GCs until the second week after challenge, when the average falls to approximately two per GC. Most significantly, the canonical V186.2 gene segment does not predominate in secondary GCs. Instead, C1H4, V23, and CH10 V_H genes are the most commonly recovered after day 4 of the secondary response. Initially, the V_H genes recovered from secondary GCs have no, or few (between one and two) mutations; de novo V(D)J hypermutation proceeds with kinetics that are similar to that observed in primary GC B cell populations (Jacob et al. 1993)

This study demonstrates that at least in the anti-NP response of C57BL/6 mice, secondary GCs do not support large populations of memory B cells that express mutated, canonical rearrangements. Our observations provide an explanation for the study of Siekevitz et al. (1987) which concluded that transferred memory B lymphocytes did not acquire new mutations upon rechallenge.

The absence of memory B cells in secondary GCs may be unique to the $\lambda 1^+$, NP-specific cells of *Igh*^{b} mice. Perhaps new rounds of mutation rapidly debilitate memory cells that reenter the GC compartment. Nonetheless, these data do not support the notion that memory B cells constitute a distinct lineage of lymphocytes that are renewed by proliferation in GCs (Linton et al. 1989, 1991, 1992; Decker et al. 1995; Klinman 1996).

Recruitment of non-canonical VDJ rearrangements into secondary GCs depends upon the presence of circulating antibody and complement. Presumably, these abundant but lower affinity (DAL PORTO et al., to be published), B cells are activated and persist within GCs by virtue of increased signaling through the CD19/CD21/CD81 complex (AHEARN et al. 1996; CROIX et al. 1996; FISCHER et al. 1996; STRAUS-SCHOENBERGER et al. 1996). Indeed, passive administration of NP-specific IgG before primary immunization with NP-CG abrogates the usual dominance of V186.2 rearrangements in GCs (CHEN and KELSOE, unpublished data).

6 Prospectus

Where do we go from here? The key issue in understanding the post-rearrangement diversification of V(D)J is the elucidation of the molecular bases of hypermutation and gene conversion. At the moment, we are only sure what the agents of these processes are not (TEXIDO et al. 1996). Perhaps the most promising advance in this problem has been the recent establishment of several in vitro systems for Ig mutation that appear to recapitulate many aspects of authentic V(D)J hypermutation (GREEN et al. 1995; ZHU et al. 1996; DÉNEPOUX et al. 1997).

The relationship between programmed V(D)J hypermutation and gene conversion also merits further close examination. Conversion has often been postulated as a potential mechanism for hypermutation and just as frequently declared wholly irrelevant (FORD et al. 1994b). Nonetheless, it is remarkable that antigen-induced GCs in chickens support both mechanisms of diversification in two distinct waves: diversification by gene conversion is followed by hypermutation (ARAKAWA et al. 1996; H. YAMAGISHI, personal communication). This echoes V(D)J mutation and/or conversion in gut-associated lymphoid follicles and reinforces the idea that both mechanisms may share some component(s) (KELSOE 1995, 1996).

The role of GCs, or more generally, the role of specific follicular architecture in inducing, maintaining, and/or catalyzing V(D)J hypermutation needs to be clarified. If hypermutation can be induced by relatively simple signals in the absence of environmental cues, powerful in vitro and biochemical studies will quickly follow. In any event, the function of the FDC is a crucial point; it may be significant that histologically similar cells are present in the gut-associated follicles of those species that exhibit developmentally programmed Ig diversification by mutation and gene conversion.

Components of innate immunity, especially the complement system, regulate the sensitivity and magnitude of humoral immune responses. In fact, activated complement proteins (e.g., C3d) and their cellular receptors (CD21, CD35) control GC activity (AHEARN et al. 1996; CROIX et al. 1996; FISCHER et al. 1996; STRAUS-SCHOENBERGER et al. 1996). Much of this control is likely determined indirectly by synergy between BCR and CD19 signals. On the other hand, we have recently shown that CD21 is necessary for the survival of even those GC B cells that exhibit extraordinarily high affinities ($> 10^{10}$ M^{-1}) for antigen (FISCHER et al., to be published).

Hypermutation in GC T cells will not be accepted until lines or hybridomas containing mutated TCR a genes can be regularly prepared. The demonstration that murine GC T cells down-regulate expression of Thy-1 (ZHENG et al. 1996a) makes possible the recovery of GC T cell populations by flow cytometry and their specific analysis. Fusion of sorted GC T cells with transformed cell lines constitutively expressing high levels of the anti-apoptotic protein, Bcl-2, should provide representative samples of the GC T cell population. We have also created transgenic mice that carry Vα and Vβ TCR rearrangements within the Lκ construct prepared by NEUBERGER and MILSTEIN's groups. Mutations introduced into this artificial substrate (BETZ et al. 1994; YÉLAMOS et al. 1995) by authentic (Ig!) hypermutation can now be compared to that recovered from single GC T cells (ZHENG 1994, 1996b).

Finally, the role(s) of RAG-1 and RAG-2 in GC B cells needs to be understood. The V(D)J recombinase may well have no important function there, merely reflecting reactivation of mechanisms to ensure self-tolerance by apoptosis (HAN et al. 1996). On the other hand, reactivation of the V(D)J recombinase may be a mechanism for self-tolerance (RADIC and ZOUALI 1996) or for the rescue of GC cells debilitated by hypermutation (HAN et al. 1997). The frequency (if any) of secondary V(D)J rearrangements in GCs must be measured and the fate of revised cells determined. To this end, transgenic recombinational substrates may prove useful. Also, the ability of RAG-1 and RAG-2 to introduce promiscuous, single-strand nicks in DNA suggests a potential role in nucleating repair-dependent mutation. Where and under what conditions are these nicks created?

There seems to be enough to keep us all busy.

References

Ahearn JM, Fischer MB, Croix D, Goerg S, Ma M, Xia J, Zhou X, Rothstein TL, Carroll MC (1996) Disruption of the Cr2 locus results in a reduction in B-1a cells and in an impaired B cell response to T-dependent antigen. Immunity 4:251–264

Allen D, Simon T, Sablitzky F, Rajewsky K, Cumano A (1988) Antibody engineering for the analysis of affinity maturation of an anti-hapten response. EMBO J 7:1995–2001

Arakawa H, Furusawa S, Ekino S, Yamagishi H (1996) Immunoglobulin gene hyperconversion ongoing in the chicken splenic germinal centers. EMBO J 15:2540–2546

Archer OK, Sutherland DER, Good RA (1963) Appendix of the rabbit: a homologue of the bursa in the chicken? Nature 200:337–339

Bachl J, Wabl M (1995) Do T-cells hypermutate? Nature 375:285–286

Banks TA, Rouse BT, Kerley MK, et al (1995) Lymphotoxin-a-deficient mice. Effects on secondary lymphoid organ development and humoral immune responsiveness. J Immunol 155:1685–1693

Becker RS, Knight KL (1990) Somatic diversification of immunoglobulin heavy chain VDJ genes: evidence for somatic gene conversion in rabbits. Cell 63:987–997

Berek C, Milstein C (1987) Mutation drift and repertoire shift in the maturation of the immune response. Immunol Rev 96:23–41

Berek C, Berger A, Apel M (1991) Maturation of the immune response in germinal centers. Cell 67:1121–1129

Berek C, Ziegner M (1993) The maturation of the immune response. Immunol Today 14:400–404
Betz AG, Neuberger MS, Milstein C (1993a) Discriminating intrinsic and antigen-selected mutational hotspots in immunoglobulin V genes. Immunol Today 14:405–411
Betz AG, Rada C, Pannell R, Milstein C, Neuberger MS (1993b) Passenger transgenes reveal intrinsic specificity of the antibody hypermutation mechanism: clustering, polarity, and specific hot spots. Proc Natl Acad Sci USA 90:2385–2388
Betz AG, Milstein C, Gonzalez-Fernandez A, Pannell R, Larson T, Neuberger MS (1994) Elements regulating somatic hypermutation of an immunoglobulin kappa gene: critical role for the intron enhancer/matrix attachment region. Cell 77:239–248
Blier PR, Bothwell A (1987) A limited number of B cell lineages generates the heterogeneity of a secondary immune response. J Immunol 139:3996–4006
Blier PR, Bothwell ALM (1988) The immune response to the hapten NP in C57BL/6 mice: insights into the structure of the B cell repertoire. Immunol Rev 105:27–43
Bothwell ALM, Pasking M, Reth M, Imanishi-Kari T, Rajewsky K, Baltimore D (1981) Heavy chain variable region contribution to the NP^b family of antibodies: somatic mutation evident in a g2a variable region. Cell 24:625–637
Carsetti R, Köhler G, Lamers M (1995) Transitional B cells are the target of negative selection in the B cell compartment. J Exp Med 181:2129–2140
Casson LP, Manser T (1995a) Evaluation of loss and change of specificity resulting from random mutagenesis of an antibody VH region. J Immunol 155:5647–5654
Casson LP, Manser T (1995b) Random mutagenesis of two complementarity determining region amino acids yields an unexpectedly high frequency of antibodies with increased affinity for both cognate antigen and autoantigen. J Exp Med 182:743–750
Chen C, Roberts VA, Rittenberg MB (1992) Generation and analysis of random point mutations in an antibody CDR2 sequence: many mutated antibodies lose their ability to bind antigen. J Exp Med 176:855–866
Chen C, Roberts VA, Stevens S, Brown M, Stenzel-Poore MP, Rittenberg MB (1995) Enhancement and destruction of antibody function by somatic mutation: unequal occurrence is controlled by V gene combinatorial associations. EMBO J 14:2784–2794
Crews S, Griffin J, Huang H, Calame K, Hood L (1981) A single VH gene segment encodes the immune response to phosphorylcholine: somatic mutation is correlated with the class of the antibody. Cell 25:59–66
Croix D, Ahearn JM, Roseng AM, Han S, Kelsoe G, Ma M, Carroll M (1996) Antibody response to a T-dependent antigen requires B cell expression of complement receptors. J Exp Med 183:1857–1864
Decker DJ, Linton PJ, Jacobs SN, Biery M, Gingeras TR, Klinman NR (1995) Defining subsets of naive and memory B cells based on their ability to somatically mutate in vitro. Immunity 2:195–203
Dénepoux S, Razanajoana D, Blanchard D, Meffre G, Capra JD, Banchereau J, Lebecque S (1997) Induction of somatic mutation in a human B cell line in vitro. Immunity 6:35–46
De Togni P, Goellner J, Ruddle NH, et al (1994) Abnormal development of peripheral lymphoid organs in mice deficient in lymphotoxin. Science 264:703–707
Diamond B, Scharff MD (1984) Somatic mutation of the T15 heavy chain gives rise to an antibody with autoantibody specificity. Proc Natl Acad Sci USA 81:5841–5844
Ferguson SE, Han S, Kelsoe G, Thompson CB (1996) CD28 is required for germinal center formation. J Immunol 156:4576–4581
Fischer MB, Ma M, Goerg S, Zhou X, Finco O, Han S, Kelsoe G, Rothstein T, Kummer E, Rosen FS, Carroll M (1996) Regulation of B cell responses to a T-dependent antigen by the classical pathway of complement. J Immunol 157:549–556
Ford JE, Mcheyzer-Williams MG, Lieber MR (1994a) Chimeric molecules created by gene amplification interfere with the analysis of somatic hypermutation of murine immunoglobulin genes. Gene 142:279–283
Ford JE, Mcheyzer-Williams MG, Lieber MR (1994b) Analysis of individual immunoglobulin lambda light chain genes amplified from single cells is inconsistent with variable region gene conversion in germinal-center B cell somatic mutation. Eur J Immunol 24:1816–1822
Fuller KA, Kanagawa O, Nahm MH (1993) T cells within germinal centers are specific for the immunizing antigen. J Immunol 151:4505–4512
Fuschiotti P, Fitts MG, Pospisil R, Weinstein PD, Mage RG (1997) RAG-1 and RAG-2 in developing rabbit appendix subpopulations. J Immunol 158:55–64
Gay D, Saunders T, Camper S, Weigert M (1993) Receptor editing: an approach by autoreactive B cells to escape tolerance. J Exp Med 177:999–1008

Green NS, Rabinowitz JL, Zhu M, Kobrin BJ, Scharff MD (1995) Immunoglobulin variable region hypermutation in hybrids derived from a pre-B- and a myeloma cell line. Proc Natl Acad Sci USA 92:6304–6308
Greenberg AS, Arila D, Hughes M, Hughes A, McKinney EC, Flajnik MF (1995) A new antigen receptor gene family that undergoes rearrangement and extensive somatic diversification in sharks. Nature 374:168–173
Griffiths GM, Berek C, Kaartinen M, Milstein C (1984) Somatic mutation and the maturation of immune response to 2-phenyl oxazolone. Nature 312:271–275
Gu H, Tarlinton D, Müller W, Rajewsky K, Förster I (1991) Most peripheral B cells in mice are ligand selected. J Exp Med 173:1357–1371
Han S, Hathcock K, Zheng B, Kepler T, Hodes R, Kelsoe G (1995a) Cellular interaction in germinal centers. The roles of CD40-ligand and B7-2 in established germinal centers. J Immunol 155:556–567
Han S, Zheng B, Dal Porto J, Kelsoe G (1995b) In situ studies of the primary immune response to (4-hydroxy-3-nitrophenyl) acetyl. IV. Affinity-dependent, antigen-driven B cell apoptosis in germinal centers as a mechanism for maintaining self tolerance. J Exp Med 182:1635–1644
Han S, Zheng B, Schatz DG, Spanopoulou E, Kelsoe G (1996) Neoteny in lymphocytes: Rag-1 and Rag-2 expression in germinal center B cells. Science 274:2094–2097
Han S, Dillon SR, Zheng B, Shimoda M, Schlissel MS, Kelsoe G (1997) V(D)J recombinase activity in a subset of germinal center B lymphocytes. Science 278:301–305
Han S, Zheng B, Takahashi Y, Kelsoe G (1997) Distinctive characteristics of germinal center B cells. Semin Immunol 9 (to be published)
Hardy RR, Carmack CE, Shinton SA, Kemp JD, Haya Kawa K (1991) Resolution and characterization of pro-B and pre-pro-B cell stages in normal mouse bone marrow. J Exp Med 173:1213–1225
Hikida M, Mori M, Takai T, Tomochika K, Hamatani K, Ohmori H (1996) Reexpression of Rag-1 and Rag-2 genes in activated mature mouse B cells. Science 274:2092–2094
Hinds-Frey KR, Nishikata H, Litman RT, Litman GW (1993) Somatic variation precedes extensive diversification of germline sequences and combinatorial joining in the evolution of immunoglobulin heavy chain diversity. J Exp Med 178:815–824
Jacob J, Kassir R, Kelsoe G (1991a) In situ studies of the primary immune response to (4-hydroxy-3-nitrophenyl) acetyl. I. The architecture and dynamics of responding cell populations. J Exp Med 173:1165–1175
Jacob J, Kelsoe G, Rajewsky K, Weiss U (1991b) Intraclonal generation of antibody mutants in germinal centers. Nature 354:389–392
Jacob J, Kelsoe G (1992) In situ studies of the primary immune response to (4-hydroxy-3-nitrophenyl) acetyl. II. A common clonal origin for periarteriolar lymphoid sheath-associated foci and germinal centers. J Exp Med 176:679–687
Jacob J, Przylepa J, Miller C, Kelsoe G (1993) In situ studies of the primary immune response to (4-hydroxy-3-nitrophenyl) acetyl. III. The kinetics of V-region mutation and selection in germinal center B cells. J Exp Med 178:1293–1307
Kelsoe G (1995) In situ studies of the germinal center reaction. Adv Immunol 60:267–288
Kelsoe G (1996) Life and death in germinal centers (redux). Immunity 4:107–110
Kelsoe G, Zheng B (1993) Sites of B-cell activation in vivo. Curr Opin Immunol 5:418–422
Kelsoe G, Zheng B, Kepler TB (1995) Do T-cells hypermutate? Nature 375:286 (reply to Bachl and Wabl)
Kimoto H, Nagaoka H, Adachi Y, et al (1997) Accumulation of somatic hypermutation and antigen-driven selection in rapidly cycling surface Ig^+ germinal center (GC) B cells which occupy GC at high frequency during the primary anti-hapten response in mice. Eur J Immunol 27:268–279
Klinman NR (1996) In vitro analysis of the generation and propagation of memory B cells. Immunol Rev 150:91–111
Knight KL, Crane MA (1995) Development of the antibody repertoire in rabbits. Ann N Y Acad Sci 764:198–206
Kolchanov NA, Solovyov VV, Rogozin IB (1987) Peculiarities of immunoglobulingene structures as a basis for somatic mutation emergence. FEBS Lett 214:87–91
Linton PJ, Decker DJ, Klinman NR (1989) Primary antibody forming cells and secondary B cells are generated from separate precursor cell subpopulations. Cell 59:1049–1059
Linton PJ, Rudie A, Klinman NR (1991) Tolerance susceptibility of newly generating memory B cells. J Immunol 146:4099–4104
Linton PJ, Lo D, Lai L, Thorbecke GJ, Klinman NR (1992) Among naive precursor cell subpopulations only progenitors of memory B cells originate germinal centers. Eur J Immunol 22:1293–1297
Liu YJ, Joshua DE, Williams GT, Smith CA, Gordon J, MacLennan ICM (1989) Mechanisms of antigen-driven selection in germinal centres. Nature 342:929–931

Liu YJ, Zhang J, Lane PJ, Chan EY, MacLennan ICM (1991b) Sites of specific B cell activation in primary and secondary responses to T cell-dependent and T cell-independent antigens. Eur J Immunol 21:2951–2962
Liu YJ, Mason DY, Johnson GD, Abbot S, Gregory CD, Hardie DL, Gordon J, MacLennan ICM (1991a) Germinal center cells express bcl-2 protein after activation by signals which prevent their entry into apoptosis. Eur J Immunol 21:1905–1910
Liu ZG, Smith SW, McLaughlin KA, Schwartz LM, Osborne BA (1994) Apoptotic signals delivered through the T-cell receptor of a T-cell hybrid require the immediate early gene Nur77. Nature 367:281–284
MacLennan ICM (1994) Germinal centers. Annu Rev Immunol 12:117–139
Malipiero UV, Levy NS, Gearhart PJ (1987) Somatic mutation in antiphosphorylcholine antibodies. Immunol Rev 96:59–74
Manser T, Huang SY, Gefter M (1984) Influence of clonal selection on the expression of immunoglobulin variable region genes. Science 226:1283–1288
Manser T, Wysocki LJ, Margolies MN, Gefter ML (1987) Evolution of antibody variable region structure during the immune response. Immunol Rev 96:141–162
Matsumoto M, Lo SF, Carruthers CJ, Min J, Mariathasan S, Huang G, Plas DR, Martin SM, Geha RS, Nahm MN, Chaplin DD (1996) Affinity maturation without germinal centres in lymphotoxin-alpha-deficient mice. Nature 382:462–466
McHeyzer-Williams MG, Nossal GJ, Lalor PA (1991) Molecular characterization of single memory B cells. Nature 350:502–505
McHeyzer-Williams MG, McLean MJ, Lalor PA, Nossal GJV (1993) Antigen-driven B cell Differentiation in vivo. J Exp Med 178:295–307
McHeyzer-Williams MG, Davis MM (1995) Antigen-specific development of primary and memory T cells in vivo. Science 268:106–111
Miller C, Kelsoe G (1995) Immunoglobulin V_H hypermutation is absent in the germinal centers of aged mice. J Immunol 155:3377–3384
Miller C, Stedra J, Kelsoe G, Cerny J (1995) Facultative role of germinal centers and T cells in the somatic diversification of IgV_H genes. J Exp Med 181:1319–1331
Monroe JG (1996) Tolerance sensitivity of immature-stage B cells. Can developmentally regulated B cell antigen receptor (BCR) signal transduction play a role? J Immunol 156:2657–2666
Nagata S, Golstein P (1995) The Fas death factor. Science 267:1449–1456
Nemazee DA, Bürki K (1989) Clonal deletion of B lymphocytes in a transgenic mouse bearing anti-MHC class 1 antibody genes. Nature 337:562–566
Pascual V, Liu YJ, Magalski A, de Bouteiller O, Banchereau J, Capra JD (1994) Analysis of somatic mutation in five B cell subsets of human tonsil. J Exp Med 180:329–339
Pulendran B, Kannourakis G, Nouri S, Smith KGC, Nossal GJV (1995) Soluble antigen can cause enhanced apoptosis of germinal-centre B cells. Nature 375:331–334
Radic MZ, Erickson J, Litwin S, Weigert M (1993) B lymphocytes may escape tolerance by revising their antigen receptors. J Exp Med 177:1165–1173
Radic MZ, Zouali M (1996) Receptor editing, immune diversification, and self-tolerance. Immunity 5:505–511
Rajewsky K, Forster L, Cumano A (1987) Evolutionary and somatic selection of the antibody repertoire in the mouse. Science 238:1088–1094
Rajewsky K (1996) Clonal selection and learning in the antibody system. Nature 381:751–758
Reynaud CA, Anquez V, Grimal H, Weill JC (1987) A hyperconversion mechanism generates the chicken light chain preimmune repertoire. Cell 48:379–388
Reynaud CA, Bertocci B, Dahan A, Weill JC (1994) Formation of the chicken B-cell repertoire: ontogenesis, regulation of Ig gene rearrangement, and diversification by gene conversion. Adv Immunol 57:353–378
Reynaud CA, Garcia C, Hein WR, Weill JC (1995) Hypermutation generating the sheep immunoglobulin repertoire is an antigen-independent process. Cell 80:115–125
Rogozin IB, Kolchanov NA (1992) Somatic hypermutagenesis in immunoglobulin genes. II. Influence of neighbouring base sequences on mutagenesis. Biochim Biophys Acta 1171:11–18
Rose ML, Birbeck MS, Wallis VJ, Forrester JA, Davies AJ (1980) Peanut lectin binding properties of germinal centres of mouse lymphoid tissue. Nature 284:364–366
Schroder AE, Greiner A, Seyfert C, Berek C (1996) Differentiation of B cells in the nonlymphoid tissue of the synovial membrane of patients with rheumatoid arthritis. Proc Natl Acad Sci USA 93:221–225

Sharon J, Gefter ML, Wysocki LJ, Margolies MN (1989) Recurrent somatic mutations in mouse antibodies to p-azophenylarsonate increase affinity for hapten. J Immunol 142:596–601
Shlomchik MJ, Marshak-Rothstein A, Wolfowicz CB. Rothstein TL, Weigert MG (1987) The role of clonal selection and somatic mutation in autoimmunity. Nature 328:805–811
Shokat KM, Goodnow CC (1995) Antigen-induced B-cell death and elimination during germinal-centre immune responses. Nature 375:334–338
Siekevitz M, Kocks C, Rajewsky K, Dildrop R (1987) Analysis of somatic mutation and class switching in naive and memory B cells generating adoptive primary and secondary responses. Cell 48:757–770
Smith KGC, Hewitson TD, Nossal GJV, Tarlinton DM (1996) The phenotype and fate of antibody-forming cells of the splenic foci. Eur J Immunol 26:444–448
Straus-Schoenberger J, Karr R, Chaplin D (1996) Markedly impaired humoral immune response in mice decifient in complement receptor 1 and 2. Proc Natl Acad Sci USA 93:3357–3361
Tew JG, Kosco MH, Burton GF, Szakal AK (1990) Follicular dendritic cells as accessory cells. Immunol Rev 117:185–211
Texido G, Jacobs H, Meiering M, Kuhn R, Roes J, Muller W, Gilfillan T, Fugiwara H, Kikutani H, Yoshida N, Amakura R, Benoist C, Mathis D, Kishimoto T, Mak TW, Rajewsky K (1996) Somatic hypermutation occurs in B cells of terminal deoxynucleotidyl transferase-, CD23-, interleukin-4-, IgD-, and CD30-deficient mouse mutants. Eur J Immunol 26:1966–1969
Thompson CB, Neiman PE (1987) Somatic diversification of the chicken immunoglobulin light chain gene is limited to the rearranged variable gene segment. Cell 48:369–378
Tiegs SL, Russell DM, Nemazee D (1993) Receptor editing in self-reactive bone marrow B cells. J Exp Med 177:1009–1620
Valitutti S, Müller S, Cella M, Padovan E, Lanzavecchia A (1995) Serial triggering of many T-cell receptors by a few peptide-MHC complexes. Nature 375:148–151
Van Rooijen N, Claasen E, Eikelenboom P (1996) Is there a single differentiation pathway for all antibody-forming cells in the spleen? Immunol Today 7:193–195
Vora KA, Manser T (1995) Altering the antibody repertoire via transgene homologous recombination: evidence for global and clone-autonomous regulation of antigen-driven B cell differentiation. J Exp Med 181:271–281
Wagner SD, Milstein C, Neuberger MS (1995) Codon bias targets mutation. Nature 376:732
Wagner SD, Neuberger MS (1996) Somatic hypermutation of immunoglobulin genes. Annu Rev Immunol 14:441–458
Weigert MG, Cesari IM, Yonkovich SJ, Cohn M (1970) Variability in the lambda light chain sequences of mouse antibody. Nature 228:1045–1047
Weinstein PD, Anderson AO, Mage RG (1994) Rabbit IgH sequences in appendix germinal centers: V_H diversification by gene conversion-like and hypermutation mechanisms. Immunity 1:647–659
Weiss U, Rajewsky K (1990) The repertoire of somatic antibody mutants accumulating in the memory compartment after primary immunization is restricted through affinity maturation and mirrors that expressed in the secondary response. J Exp Med 172:1681–1689
Wilson M, Hsu E, Marcuz A, Courtet M, Du Pasquier L, Steinberg C (1992) What limits affinity maturation of antibodies in Xenopus, the rate of somatic mutation or the ability to select mutants? EMBO J 11:4337–4347
Woronicz JD, Calnan B, Ngo V, Winoto A (1994) Requirement for the orphan steroid receptor Nur77 in apoptosis of T-cell hybridomas. Nature 67:277–281
Wysocki L, Manser T, Gefter ML (1986) Somatic evolution of variable region structures during an immune response. Proc Natl Acad Sci USA 83:1847–1851
Yang X, Stedra J, Cerny J (1996) Relative contribution of T and B cells to hypermutation and selection of the antibody repertoire in germinal centers of aged mice. J Exp Med 183:959–970
Yélamos J, Klix N, Goyenechea B, et al (1995) Targeting of non-Ig sequences in place of the V segment by somatic hypermutation. Nature 376:225–229
Zheng B, Xue W, Kelsoe G (1994) Locus-specific somatic hypermutation in germinal centre T cells. Nature 372:556–559
Zheng B, Han S, Kelsoe G (1996a) T helper cells in murine germinal centers are antigen-specific emigrants that down-regulate Thy-1. J Exp Med 184:1083–1091
Zheng B, Han S, Zhu Q, Goldsby R, Kelsoe G (1996b) Alternative pathways for the antigen-specific selection of peripheral T cells. Nature 384:263–266
Zhu M, Green NS, Rabinowitz JL, Scharff MD (1996) Differential V region mutation of two transfected Ig genes and their interaction in cultured B cell lines. EMBO J 15:2738–2747

Somatic Mutagenesis and Evolution of Memory B Cells

L.J. WYSOCKI[1,3], A.H. LIU[2,4], and P.K. JENA[1]

1 Introduction

The parallels between classical evolution of species and somatic evolution of lymphocytes are astounding. While ontogeny is sometimes an artful rendition of phylogeny, somatic evolution in vertebrate immunity is an accelerated version of the evolutionary algorithm that shaped complex life forms on this planet. Somatic evolution has adopted cardinal features of classical evolution, even at a high level of resolution. If "survival of the fittest" is the onus of life, then "survival of the best fit" appears to be the onus of the B lymphocyte and its antibody product.

[1]Division of Basic Sciences, Department of Pediatrics, National Jewish Medical and Research Center, 1400 Jackson Street, Denver, Colorado 80206, USA
[2]Division of Allergy, Department of Pediatrics, National Jewish Medical and Research Center, 1400 Jackson Street, Denver, Colorado 80206, USA
[3]Department of Immunology, University of Colorado Health Sciences Center, Denver, Colorado 80262, USA
[4]Department of Pediatrics, University of Colorado Health Sciences Center, Denver, Colorado 80262, USA

Mutation, selection, and propagation of mutations describe the most basic algorithm of classical evolution. It is likewise the overriding theme for development and differentiation of B and T lymphocytes, which utilize three forms of mutagenesis, and which are subject to numerous and often unforgiving measures of selection. While differences in the form of somatic mutation utilized and the quality of selection pressure faced can be found between B and T lymphocytes and among lymphocytes of different vertebrates, the overall similarities are striking. Exon shuffling, gene conversion and point mutagenesis create lymphocyte receptor diversity in vertebrates, while lymphocyte survival and propagation are determined by selection pressures that usually reward foreign reaction, and punish self reaction.

Lymphocytes of the preimmune repertoire evolve by major mutagenic events that build the genes for their antigen receptors that, in turn, render each cell unique to a first approximation. In some species, receptor variable (V) genes assemble from smaller exon-like segments, where numerous alternative combinations of gene segments are available for a given lymphocyte, as in mouse and man (TONEGAWA 1983). This somatic movement and assembly of gene segments bears a striking similarity to the exon shuffling events that GILBERT (1978) has proposed to occur in the evolution of the eukaryotic genome. In other species such as chickens and rabbits, much of this combinatorial diversity is substituted by a gene conversion process, which bears a striking resemblance to meiotic conversion events that also molded the genome (REYNAUD et al. 1987; THOMPSON and NEIMAN 1987; KNIGHT and CRANE 1995). Still other species exploit massive point mutation to build preimmune receptor diversity in the lymphocyte repertoire, as shown so elegantly in sheep by REYNAUD et al. (1995).

Nowhere is the parallel with meiotic evolution so evident as in the B lymphocytes of mice and humans, where preimmune diversity is amplified by a postimmune diversification mechanism variously referred to as somatic mutation, somatic mutagenesis and somatic (hyper)mutation (WEIGERT et al. 1970; CREWS et al. 1981; reviewed by FRENCH et al. 1989). The enormous preimmune antibody repertoire of B cells in these species is generated by the V gene segment assembly process. Approximately 10^7 distinct combinations of V gene segments can encode the heavy and light chain V regions of antibodies. And this diversity is amplified by substantial heterogeneity created by untemplated polymerase and nuclease activities at gene segment joints (DESIDARIO et al. 1984; LIEBER 1996). Preimmune diversity is greatest at segment joints, and segment joints encode a portion of the V region polypeptide (CDR3) that is very likely to directly engage a foreign antigen. Based on first principles, it would appear that the potential of this preimmune diversity of antibodies, which is derived from V segment combinations and junctional variation, exceeds the number of lymphocytes present in the animal at any given time. Yet, despite this seemingly comprehensive repertoire of preimmune antibody structures, the B lymphocyte receptor of mice and humans is subjected to somatic diversification once more, following recruitment into the immune response by antigen.

Antigen-driven B cell development in mice and humans is a second phase of somatic evolution. At this time, a fascinating but poorly understood somatic mutation process introduces nucleotide changes into antibody V genes. An equally impressive selection process directs the survival and proliferation of B lymphocytes expressing receptor antibody with improved fit for antigen. The receptor antibody is the principal target of selection pressure exerted on the B cell. It decides on life or death. It is the beak of a Darwin finch.

This chapter will focus on the most basic features of this somatic evolutionary process as they pertain to the development of memory B cells in mice and humans. We will focus on major current questions regarding selection pressures faced by developing memory B cells, population dynamics of memory progenitors, and somatic mutagenesis in antibody V genes. Given the short space, the chapter is not intended to be comprehensive. We recommend recent complementary reviews on related aspects of B cell memory (MACLENNAN 1994; KELSOE 1996a; KLINMAN 1996; RAJEWSKY 1996).

2 Selection Pressures That Drive Development of Memory B Cells

2.1 The Power of Positive Selection by Antigen

There is universal agreement that B lymphocytes are subjected to intense selection pressures during antigen-driven development. Even without pretense to the nature of the selection pressures, this conclusion can be deduced by the observation that secondary immune responses tend to be dominated by progeny of relatively few B cell precursors that have undergone massive clonal expansion (MCKEAN et al. 1984; CLARKE et al. 1985; BLIER and BOTHWELL 1987; CLAFLIN et al. 1987). The same kinds of hybridoma sampling studies that revealed this also revealed that members of an expanded B cell clone, though mutationally diverse, often share specific somatic mutations and sometimes combinations of mutations. It is inferred that some of these shared mutations, marking branchpoints within phylogenetic clonal trees, exerted a positive influence on B cell selection and propagation.

A complementary body of data, derived almost entirely from analyses of anti-hapten responses, has indicated the importance of antibody affinity to B selection, and especially to selection of B cells expressing mutant receptor antibodies. Almost without exception, mutant hybridoma antibodies of the secondary immune response bind the eliciting hapten with improved affinity relative to their non-mutant predecessors (GRIFFITHS et al. 1984; WYSOCKI et al. 1986; MALIPIERO et al. 1987). The power of anti-hapten studies derives from their reproducibility and from the static nature of the antigen. We have exploited the strain A immune response to *p*-azophenylarsonate (Ars), because a highly-defined antibody is consistently elicited in the memory immune response (reviewed in MANSER et al. 1987). It is a marker for memory B cells, and we refer to this antibody, the V genes that encode it, and the B cells that synthesize it as "canonical". Canonical anti-Ars antibodies are encoded by one set of V_H, D, J_H, V_κ, and J_κ gene segments, with junctional diversity limited to amino acids 100 and 107 in CDR3 of the heavy chain (WYSOCKI et al. 1987). For the sake of clarity, note that there are many analogues of *nearly* canonical anti-Ars antibodies elicited during the response distinguishable by D or J_H segment usage or sometimes only by the number of junctional codons at segment boundaries. In fact, these analogues are more frequent than the canonical element during early stages of the primary immune response (Wysocki et al. 1986). The canonical element is simply the most operationally useful, because it is the most consistent antibody structure of the secondary (memory) immune response, where it comprises, on average but with some variation, approximately 40% of all anti-Ars antibodies. Accordingly, it is the highest-resolution structural definition of recurrent B cell memory available.

The structural basis of affinity improvement is the somatic mutation, and virtually all anti-hapten systems examined are characterized by recurrent somatic mutations that confer affinity improvements (Berek and Milstein 1987; Allen et al. 1988; Sharon et al. 1989). In the Ars model, these are isoleucine and threonine replacements at residues 58 and 59 in V_H CDR2 (Sharon et al. 1989). Approximately one half of secondary canonical B cells express one or both of these somatic mutations, presumably due to their positive influences on B cell proliferation. If an iodinated form of Ars is used to immunize strain A mice, a new recurrent mutation is observed that improves affinity for this modified hapten (Fish et al. 1991). Similarly, Lozano et al. (1993) have shown that a selection pressure favors B cells that preferentially express one particular copy of a κ transgene with a beneficial somatic mutation in a mouse that carries five copies of the transgene. Expression of the other four copies lacking the affinity-enhancing mutation would presumably diminish receptor occupancy by antigen (oxazalone) on the B cell surface, and nonsense mutations have inactivated most of the other copies in the sampled B cell clones. Since most randomly-occurring mutations do not improve affinity for a particular antigen, the recurrence of specific affinity-improving mutations thus provides compelling evidence for the importance of affinity, and presumably receptor occupancy, in the B cell selection process (Chen et al. 1992; Casson and Manser 1995a).

The consequences of this apparent selection pressure for receptor occupancy are sometimes remarkable. Four monoclonal anti-Ars antibodies have been identified that share a cluster of four amino acid replacements out of eight V_H CDR2 residues (Wysocki et al. 1990) (Fig. 1). The cluster of mutations improve affinity for Ars (Parhami-Seren et al. 1990). These antibodies are products of independent cell fusions performed in three laboratories (Slaughter and Capra 1983; Fish et al. 1989; Wysocki et al. 1990). The alterations are almost certainly the consequence of accumulated point mutations, as indicated by the absence of a potential donor sequence in the A/J genome, by the expression of one or more of these mutations in various combinations by other hybridomas, and by the use of alternative nucleotide substitutions for one of the amino acid replacements. This is a striking example of convergent evolution that we are compelled to believe is driven, at least in part, by receptor occupancy (with antigen) constraints.

One of the most striking cases illustrating the adaptive power of somatic mutagenesis and selection comes from studies of transgenic mice that carry between

	V_H CDR2									
	50									60
Germline	I	N	P	G	N	G	Y	T	K	Y
mAb LA2	-	H	-	-	K	-	-	I	H	-
mAb 91A3	-	H	-	-	K	-	-	I	H	-
mAb 45-49	-	H	-	-	K	-	-	I	H	-
mAb 8AS2A	-	H	-	-	K	-	-	I	H	-

Fig. 1. Convergent evolution of antibody structure. The four monoclonal antibodies were derived from *p*-azophenylarsonate (Ars)-immune mice in independent fusions performed in different laboratories [91A3: Slaughter and Capra (1983); 45–49 and LA2: Wysocki et al. (1990) and unpublished data; 8AS2A: Fish et al. (1989)]. Each expresses the Ars-associated canonical V_H gene. Their CDR2 sequences are compared to that of the unmutated canonical V_H at the top. Mutations that produce amino acid replacements are explicitly indicated

30 and 50 copies of a canonical μ transgene. Durdik et al. (1989) have shown that these mice produce transgene-derived canonical anti-Ars antibodies upon immunization. We have used this mouse to test the adaptive power of somatic evolution during immunity.

The transgenic experiment was conceived from an earlier experiment involving normal A/J mice, where canonical mutants with changed antigenic specificity were selected and immortalized in the form of hybridomas (Ellenberger et al. 1993). Immunization with sulfanilic acid (Sulf) does not normally elicit a detectable response by germline-encoded canonical clones. But if canonical B cells are first recruited into an immune response with Ars, somatic mutants that happen to acquire affinity for Sulf can be selected by delivering Sulf injections shortly thereafter. In acquiring affinity for Sulf, the mutants lose affinity for Ars. If the Ars-immunized animals are rested for several months before receiving the Sulf booster injections, switch specificity mutants of this type are not obtained, apparently because Sulf must be available during the period of mutational diversification in order to recruit the Sulf-binding, Ars-nonbinding mutants into the memory repertoire (Fish et al. 1989). The Sulf-binding mutants invariably contain one of several possible replacement mutations at codon 35 in V_H CDR1. In vitro mutagenesis studies have shown that these mutations are largely responsible for the switch in specificity, and this is consistent with the observation that mutations of this type, or even any replacement mutation at this codon, have never been observed in canonical antibodies elicited strictly with Ars (Kussie et al. 1994).

When a similar but exaggerated immunization strategy (six Sulf booster injections) was applied to the canonical-μ transgenic animal, a grossly exaggerated form of the same result was obtained (Jena and Wysocki, unpublished data). First, several of the resulting hybridomas expressed canonical Sulf-binding IgG antibodies, apparently due to transchromosomal isotype switching, as described by Gerstein et al. (1990) and to expression of an endogenous canonical κ chain. The γ transgene copies all carried diagnostic replacement mutations at residue 35 in CDR1, and they were riddled with mutations to an extent that we have not seen in any published hybridoma sequence. One of the resulting canonical transgene-expressing hybridomas carried over 31 nucleotide replacements in its V_H gene alone, resulting in replacement of 16% of its amino acids. The structural changes were so extensive that we would not have been able to conclude that these antibodies were mutationally-derived products had we not exploited transgenic technology. The massive structural changes were accompanied by equivalent functional changes in antigenic specificity. The predecessor antibody bound to Ars with high affinity, but to Sulf with a barely detectable affinity. In contrast, the mutants all bound Sulf with very high affinities, while their affinities for Ars were weak or nonexistent. For one mutant, the relative change in Sulf/Ars affinity ratio compared to its unmutated precursor was nearly 60 000-fold. All of this is made even more astounding by the observation that all other copies of the μ-canonical transgene in these hybridomas had been silenced by some mechanism that we still do not understand. Selection pressures had apparently favored B cells expressing the pure mutant receptor antibody with specificity for Sulf.

It appears that, in the game of B cell survival, failure is the rule and success is a rare and somatically-endowed exception. But persistent antigen-based selection ultimately results in the fixation and multiplication of rare exceptions, whereby the seemingly improbable becomes probable.

2.2 Negative Selection of Memory Progenitors

Somatic mutagenesis at any level is a potentially dangerous process. In antibody genes this would appear to be especially true, since mutagenesis can create autoreactive antibodies, as suggested by classic in vitro studies (Diamond and Scharff 1984). If autoreactive B cells arise by somatic mutagenesis during antigen-driven B cell differentiation, what is their fate?

Studies of B cell tolerance have indicated that during preimmune B cell development in vivo, a negative selection pressure operates to eliminate autoreactive B cells, or at least those B cells that react with multivalent self antigens. Whether by death, anergy, or receptor editing, the immune system appears committed to eliminating the autoreactive element functionally or physically (Nemazee 1993; Goodnow 1996; Nossal 1994; Klinman 1996). It is thus reasonable and even attractive to propose that tolerance mechanisms may also operate on emerging mutant B cells that have acquired reactivity for multivalent self during antigen-driven development. Tolerance susceptibility appears to be high in memory B cells, as shown in situ with hapten-carrier models, where B cell receptor occupancy was dissociated from T cell help (Linton et al. 1991) and in vivo, where vital costimulatory signals between B cells and presumed helper T cells had been disrupted (Han et al. 1995a). Results of additional in vivo studies, where B cell development in an environment of abnormal antigen-excess was examined, have also suggested that the cellular context of antigen recognition by B cells is critical to their survival (Nossal et al. 1993; Han et al. 1995b; Pulendran et al. 1995; Shokat and Goodnow 1995).

One of the more compelling experiments supporting the idea of tolerance in autoreactive B cell mutants was performed by Casson and Manser (1995b). They randomly mutated two codons in CDR-2 of the Ars canonical V_H gene and generated a phage display library expressing the artificial mutant V_H library in the context of the germline canonical κ chain. Many of the selected antibodies that had acquired improved affinities for Ars also acquired significant affinities for DNA. Yet these bipartisan mutants have never been observed in the hundreds of hybridomas sampled from Ars-immune animals. If B cell selection were exclusively based on improving affinity for antigen, one would have expected to find natural counterparts to these in vitro-generated mutants. Their absence implies that tolerance of self-reactive B cells occurs at the post-antigen driven stage of development. Competition between DNA and Ars could limit the carrier-specific T cell help available to the bipartisan B cell, or alternatively could limit potentially important interactions between the B cell and Ars-presenting follicular dendritic cells. However, all of this is conjecture until a formal demonstration that such mutants are generated and eliminated in vivo is provided.

2.3 Adaptive Potential of Somatic Mutagenesis and Selection

The affinity maturation of serum antibody with time after immunization appears to be largely the consequence of antigen-driven selection and the combined consequences of somatic mutagenesis and further antigen-driven selection. It is often inferred that affinity maturation plays a beneficial role in acquired immunity, particularly in the face of an avidity loss that almost certainly accompanies isotype switching from μ to γ. However, given the capacity of the immune system to select B cell mutants with changed antigenic specificity as discussed above, it is conceivable that somatic mutagenesis plays another significant physiological role. We would like to suggest that somatic mutagenesis in B cells may help protect against escape of mutating pathogens.

For reasons of history and convenience, most of our mechanistic knowledge regarding humoral immunity comes from the use of artificial non-living antigens. These are presumably static in nature, even after entering the animal. However, in one interesting study reported by Rickert et al. (1995) mutations within B cell genealogies were examined for their influences upon antibody affinity for the immunogen which, in this case, was a live influenza virus. Surprisingly, specific replacement mutations that appeared to be influential in selection, based on their positions at branchpoints in genealogical trees and by their independent presence in more than one hybridoma, did *not* enhance antibody binding to the original influenza immunogen. In fact, they reduced antibody affinity. Given a landslide of results from studies of hapten systems indicating a tight correlation between affinity improvement and successful B cell selection, this finding presents a significant paradox. The resolution of this paradox may lie in the nature of the immunogen, which was a live virus with the potential to mutate in vivo. It is possible that the analyzed hybridomas represented mutant B cell subclones that had been selected by a mutant form of the original virus. In this regard, Clarke et al. (1985) observed, in an earlier study, significant intraclonal variation in antibody fine specificity for a variety of influenza subtypes. We recognize that this hypothesis is not the only explanation for the data, but it has sufficiently important implications to warrant further investigation of antibody responses to live immunogens.

2.4 Other Potential Selection Pressures on Memory Progenitors

Diversification of antibody V region genes is most often viewed from the perspective of V region structure, specificity and affinity, and influences of these on B cell recruitment and propagation. Given the vital role of T cell help in many B cell responses, and specifically those that generate memory B cells with mutant receptor antibodies, it is possible that antibody V gene diversification has other important consequences to humoral immunity, tolerance, and autoimmunity. In part, this concern arises from pioneering studies by WEISS and BOGEN (1991), who demonstrated that B cells can process and present peptides of endogenously-synthesized antibody V regions in the context of class II MHC. Their finding is in perfect agreement with a preponderant representation among natural class II-presented peptides in B cells of those derived from abundant endogenous proteins destined for the membrane (RUDENSKY et al. 1991; CHICZ et al. 1992; MARRACK et al. 1993). It suggests a scenario in which B cells might receive chronic T cell help independently of foreign antigen. In addition, JORGENSEN et al. (1983) demonstrated that a somatically altered segment of the MOPC 315 myeloma light chain was immunogenic for class II MHC-restricted T cells in syngeneic mice, suggesting that the T cell repertoire might be tolerant of germline-encoded antibody diversity but not diversity generated by somatic mutagenesis.

EYERMAN et al. (1996) explicitly addressed this issue in the Ars model and found evidence for T cell tolerance of germline-encoded antibody diversity. In addition they found that physiologically-acquired somatic mutations created V region epitopes for class II MHC-restricted T cells, in some cases by improving or creating a binding capacity for class II MHC, as shown in competition binding studies with synthetic peptides that contained or lacked critical somatic mutations. In one case, a single conservative somatic mutation created a motif of peptides restricted by IE^k that was apparently responsible for its immunogenicity. Epitope-creating mutations were observed in framework and CDR, suggesting that all parts of the V region are potentially subjected to processing and MHC-presentation. Collectively, these results indicate that V gene somatic mutations create potentially immunogenic epitopes for class II MHC-restricted T cells, and that this occurs almost certainly within some members of a given B cell clone.

At present, we do not know the consequences of acquiring mutationally-created V region epitopes on the expressing B cell or potentially reactive T cells in vivo. Acquired tolerance is an attractive hypothesis, otherwise the potential for chronic activation of autoreactive B cells would seem to be high. Tolerance could be executed either in the T cell repertoire, or in the B cell repertoire, by selective elimination of the mutant B cell as it expands and becomes visible to V epitope-reactive T cells. The latter of these would constitute another selection pressure on B cells, in addition to the positive element for improved receptor affinity for antigen and the hypothesized negative element against receptor autoreactivity, as described above. There does not appear to be an advantage to acquiring V epitope-creating mutations in normal responses, because recurrent somatic mutations examined in a variety of hapten systems either improve receptor affinity for antigen or are located at nucleotide sequences that mutate at intrinsically high rates. Whether a counter selection pressure operates on B cells with mutationally-created class II MHC-restricted V region epitopes, however, remains to be determined.

3 Population Dynamics in Memory B Cell Development

3.1 Identity of the Mutating B Cell

The adaptive power of somatic evolution in generating memory B cells is a remarkable phenomenon, and accordingly, there is a great deal of interest in the underlying population dynamics. Based on an extensive body of data, there is universal agreement that some or all of memory B cell development takes place in the microenvironment of the germinal center (GC). These are sites of antigen deposition and extensive oligoclonal B cell division and differentiation in B cell-rich zones of secondary lymphoid tissue (NIEUWENHUIS and OPSTELTEN 1984). The topic has been recently reviewed by KELSOE (1995, 1996a,b) and MACLENNAN (1994) for those interested in further details. Here we will cover some controversial aspects of GCs that pertain explicitly to somatic mutagenesis and receptor-mediated selection. One of these is the nature of the mutating B cell.

GCs embody two fundamentally distinct types of B cells that form the cellular basis of their polarity. One is a rapidly dividing centroblast group at one pole termed the dark zone (ZHANG et al. 1988; LIU et al. 1991, 1992a). These cells appear to lack expression of immunoglobulin (Ig), but they give rise to the second population, termed centrocytes, that bear membrane Ig (mIg$^+$), undergo little if any cell division and are located within the light zone (FLIEDNER et al. 1964; BUTCHER et al. 1982; LIU et al. 1991, 1992a). It is commonly assumed that the centroblast must be the mutating B cell, if for no other reason than mutation is generally associated with DNA replication (LIU et al. 1992a; BEREK and ZIEGNER 1993; PETERS and STORB 1996). Some authors claim to have proved that the centroblast is the mutating cell (PASCUAL et al. 1994; LIU et al. 1996a). They draw their conclusion from the observed presence of somatic mutations in V genes of isolated human centroblasts. Since centroblasts give rise to centrocytes, somatic mutations in centrocyte V genes are presumed to be derived from mutating centroblast progenitors. Inexplicably, some of the same authors argue for the existence of a cyclic re-entry process, whereby a mutant centrocyte expressing an affinity-improved mutant receptor is positively selected to de-differentiate into a centroblast to begin the mutation, proliferation, and differentiation cycle once more (Fig. 2). If centrocytes can give

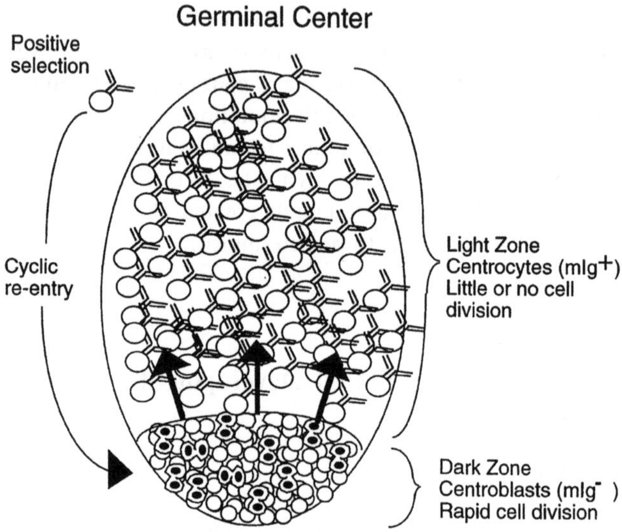

Fig. 2. The cyclic re-entry hypothesis for germinal center (GC) B cells. Dividing mIg$^-$ centroblasts give rise to mIg$^+$ centrocytes. Mutant centrocytes that are positively selected by antigen re-enter the GC as centroblasts, presumably with an accompanying loss of mIg. This generalized model does not indicate in which population somatic mutagenesis is occurring

rise to centroblasts via this cycling format, then the presence of somatic mutations in centroblasts does not indicate whether they are the mutating population. At present, we do not know which population is mutating, but there is some intriguing evidence from two avenues of investigation suggesting that the centrocyte might possibly be the mutating B cell. The first is an association between transcription and mutagenesis as discussed later (Sect. 4.3), and the second is derived from mathematical modeling studies of Kepler and Perelson (1993), suggesting that an efficient way to achieve affinity maturation is to alternate between periods of mutagenesis and periods of mutation-free growth.

3.2 To Recycle or Not

A second point of controversy, mentioned above, is whether positively-selected centrocytes de-differentiate and recycle to the centroblast compartment. The cyclic re-entry model is attractive because it fits well with genealogical data, originally obtained by Weigert et al. (1970), indicating that mutagenesis occurs through multiple cycles of DNA synthesis (McKean et al. 1984; Clarke et al. 1985). They were the first to identify the mutationally-based genealogical trees that indicated this point so clearly. It is reinforced strongly by data described above from anti-hapten studies, in which hybridomas often express antibody genes that carry multiple affinity-improving mutations in combinations. It is difficult to envision how mutation, selection, and division could occur through multiple rounds without cyclic re-entry, for dividing centroblasts, which appear to lack mIg, presumably cannot be selected, while centrocytes, which can be selected, do not appear to be dividing. Without re-entry, it would appear that only one round of mutation, selection, and division could take place in the GC.

Despite the apparent need to invoke cyclic re-entry, there are at least two alternative hypotheses to explain the hybridoma data. The first is that some centrocytes are dividing, specifically the relevant ones, those rare individuals that carry affinity improving mutations. This model requires that at least some mutagenesis occurs in centrocytes. The second model is that some mutagenesis proliferation and selection occurs outside of the GC. *Xenopus* lacks GCs, yet the antibody V genes of this species are subjected to some mutagenesis and selection, though apparently not as extensively as in mice and humans (WILSON et al. 1992). Also supporting this idea is the recent finding that lymphotoxin α-deficient mice, which apparently lack GCs, do support some somatic mutagenesis and selection of mutant B cells, though again, not as extensively as in normal GC-containing mice (MATSUMOTO et al. 1996). This conclusion should be qualified because it is formally possible that GCs may have been present in non-lymphoid tissue (SCHRODER et al. 1996). Significant proliferation of memory B cells also almost certainly occurs outside of the GC, as indicated in part from our single cell studies of Ars-binding canonical B cells with a GC phenotype, that number no more than approximately 30 000 in spleen at any one time (LIU et al. 1996b). Despite these low numbers, we observe milligram quantities of canonical serum antibody in these mice (LIU and WYSOCKI, unpublished data). Based on these findings, it is possible that some important steps in memory development, including mutagenesis, proliferation, and selection, occur outside the GC.

We have recently obtained some support for the cyclic re-entry hypothesis from single cell ex vivo studies of canonical B cells (LIU et al., unpublished data). Some of the cells appear to lack membrane Ig, yet they carry canonical V genes and sometimes with affinity-improving mutations that are indicative of positive selection. The mIg^- phenotype was confirmed by an absence of cytoplasmic κ transcripts in some of these cells. It is difficult to envision how a mIg^- cell could be positively selected unless it was derived from a precursor that was mIg^+. Hence these data argue for a mIg^+ to mIg^- de-differentiation process in antigen-responsive B cells. It is unlikely that these interesting cells are apoptotic because of their large size, which is evident upon direct microscopic visualization. Furthermore, the cells stain brightly with PNA, which together with their mIg^- phenotype, suggests they are possibly GC centroblasts and almost certainly not plasma cells (ROSE et al. 1980). Collectively, these observations are consistent with the hypothesis that the isolated cells might be centroblasts that were derived from a positively selected centrocyte, expressing a mutant receptor antibody of improved affinity. Further experiments are underway to address this possibility. At present, however, conclusive evidence for the cyclic re-entry hypothesis in GCs is lacking.

3.3 Receptor Purging

The idea that B cells re-cycle between selected mIg^+ and dividing mIg^- compartments is attractive for another reason. In principle, this scenario allows a B cell to purge itself of a previous mIg product before expressing a newly mutated derivative. Otherwise, the cell might simultaneously express two successive generations of mutant receptor antibodies and their hybrid products. It appears that B lymphocytes do have specialized mechanisms for regulating Ig transcript abundance. Provocative results from Milstein's laboratory have revealed that κ transcripts with nonsense codons are recognized by nuclear-derived factors that inhibit their splicing and transport to the cytoplasm (Lozano et al. 1993; Aoufouchi et al. 1996). We have also observed evidence of unusual Ig transcript regulation in hybridomas derived from the μ-Ars transgenic mouse described in Sect. 2.1 (Jena and Wysocki, unpublished data). Several of these hybridomas that predominantly secrete Sulf-binding IgM also secrete reduced levels of Ars-binding IgM. Using a single-molecule PCR procedure, we amplified, enumerated, and directly sequenced individual μ cDNAs from these cells and observed several categories of transcripts defined by mutation patterns. But the one transcript with a specific somatic mutation at V_H CDR-1, codon 35 that changes specificity from Ars to Sulf, was present more abundantly than all of the other transcripts combined. In addition, many other μ Ars transgenes were apparently silent. It is possible that precursors to these hybridomas were able to regulate receptor expression by regulating transcript abundance in a manner consistent with the selection pressure applied by antigen (Sulf). Observations such as these make plausible the hypothesis of mIg^+ to mIg^- de-differentiation and re-cycling in concert with antigen-driven selection.

3.4 Numbers of Memory Progenitors

Studies of memory B cell development at the single cell level have provided a data set complementary to those of GCs. The single cell approach was introduced by McHeyzer-Williams et al. (1991), who followed the response of a B cell population specific for the hapten (4-hydroxy-3-nitrophenyl)acetyl NP. A major advantage of the single cell procedure is the absence of hybrid PCR products that can form when two or more differently mutated genes are amplified simultaneously, as shown by Ford et al. (1994a). This is important, because such products do not represent natural V gene sequences and because they can lead to artifactual genealogical trees in some cases.

To study a highly reproducible component of memory, we developed procedures to isolate single canonical B cells from a suspension of splenocytes derived from immunized A/J mice (Liu et al. 1996b). Isolation was achieved by staining spleen cells with an anti-idiotype that is quite specific (approximately 80%) in its recognition of the complete canonical V region structure. After sorting, the single cells were directly visualized in microdrops prior to their aspiration into a lysis buffer for PCR amplification. Canonical V genes were then amplified and directly sequenced without cloning. Since there was no precedent for single molecule amplification and direct sequencing, a considerable effort was expended in standardizing the procedure with hybridomas, especially with regard to calibrating Taq polymerase misincorporation frequencies, which were found to be negligible (approximately 1/10 000 bases of sequence) (LIU et al. 1992b; JENA et al. 1996). Presumably only misincorporations in the first cycle should be observed in the final sequence autoradiogram. Hence, observed Taq error frequencies approximated previously calculated Taq error rates, where the rate is the number of mistakes per nucleotide per synthesis cycle. Using this single cell procedure, we identified canonical B cells that had every structural feature of memory at the highest level of resolution, including precise V_H, D, J_H, V_κ and J_κ gene segment usage, somatic mutations, and specific affinity-improving somatic mutations.

Results of these single cell studies revealed several informative features of memory B cell development (LIU et al. 1996b). First, there were relatively few mIg^+ canonical memory-lineage B cells in spleen. Even at their peak, which is 13 days post-intraperitoneal immunization, they numbered only approximately 40 000 (<0.1%) in the average spleen. This is at least ten-fold less than the number of NP-binding B lymphocytes observed in spleens of NP-immunized mice and estimated from GC content (JACOB et al. 1991; MCHEYZER-WILLIAMS et al. 1993). The number of potential canonical memory progenitors was even smaller than this, however, because approximately half of them failed to bind Ars, as determined by flow cytometric analyses of cells stained with the anti-idiotype in the presence of competing Ars-bovine serum albumin (BSA). Their failure to bind Ars implies that they were recent targets of somatic mutagenesis. The low frequency of canonical cells is consistent with a second observation that the participants in a given spleen comprised few genealogies derived from an average of between approximately four and five precursor cells. (We assign lineage relationships on the basis of junctional codon sequences at positions 100 and 107 and at the third nucleotide of the serine codon at position 99.) Because canonical products comprise approximately 40% of the secondary immune response to Ars, this implies that the entire Ars-specific splenic memory compartment consists of no more than approximately 50 000 B cells at its peak. Yet serum titers of anti-Ars antibodies from boosted mice that are identically primed often reach around 1–3 mg/ml, approximately half of which is canonical – a surprising result given the few candidate cells identified. That few canonical cells are identified is consistent with hybridoma sampling studies suggesting that relatively few clones contribute to a large fraction of the B cell memory response (MCKEAN et al. 1984; CLARKE et al. 1985; BLIER and BOTHWELL 1987; CLAFLIN et al. 1987).

The paucity of observed canonical lineages implies something else. If GCs are oligoclonal, then only products of a few GCs give rise to the lion's share of memory B cells. An alternative possibility is that single canonical B cells expand into clones that seed many GCs, but studies of JACOB and KELSOE (1992) suggest that the capacity of single periarteriolar lymphoid sheath (PALS) foci of proliferating B cells to seed GCs is limited. JACOB et al. (1993) reported evidence of autonomous competition within the GC that narrows clonal diversity with time. If seeding of GCs is limited, our results would suggest another bottleneck in memory development – that the memory response is derived from a few GCs, carved out of a much larger number of failures. Therefore, studying the average GC may not necessarily be tantamount to studying memory development in its fullest sense.

Somatic mutations were evident within the canonical V genes of isolated single cells, and in some cases affinity-improving mutations at V_H CDR2 positions 58 and 59 were present, indicating some degree of post-mutational selection. Based on a calibrated Taq error frequency, we calculate that fewer than 1/70 "mutations" was a Taq polymerase error, and this is confirmed by the sequence-specific target biases that reveal the telltale signature of the somatic mutation mechanism (see below). The overall frequency of somatic mutation at day 13, when canonical cells were observed most abundantly, was a modest ca. 1%, and mutation-based genealogical trees were relatively simple, with few branchpoints and many independent stems derived from unmutated precursor cells (Fig. 3). In other words, the "trees" looked more like "bushes" (MANSER 1989). This implies that more selection and possibly more mutation was yet to occur, which is interesting because the number of observable canonical splenic B cells with a GC surface phenotype begins to wane after day 13.

In sum, the population dynamics that bring about receptor-driven evolution of B cell memory are beginning to unfold. It is clear that many of the important events take place within the microenvironment of the GC and that a large fraction of memory can be derived from just a few precursor B cells. But there is still much to be learned about the somatic diversification of the B cell receptor and its relationship to B cell memory development. We are still uncertain of the identity of the mutating B cell. It is unclear whether cycles of mutation, differentiation and de-differentiation take place in the GC and in concert with selection, whether Ig transcripts are induced and purged in alternate cycles, and even whether some aspects of memory development occur outside the GC microenvironment.

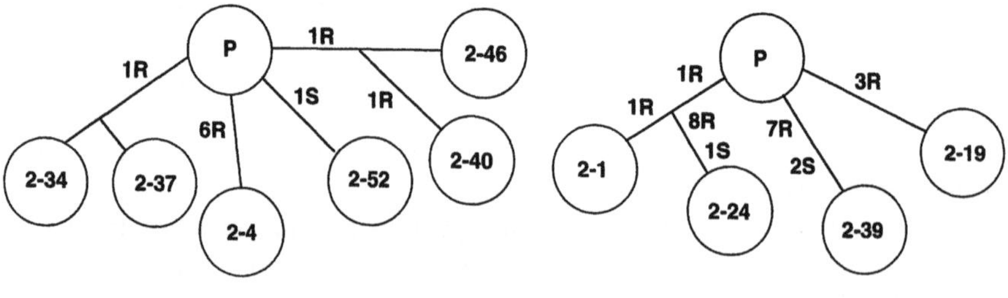

Fig. 3. Genealogical bushes. Data were obtained from V gene sequences of single splenic mIg$^+$ canonical B cells obtained 13 days after immunization. Each B cell is denoted by a number. *P*, assumed unmutated precursor cell; *R*, the number of replacement mutations; *S*, the number of silent mutations

4 Characteristics of the Somatic Mutation Mechanism

4.1 Point Mutation vs a Templated Recombinational Process

In birds and rabbits many V gene somatic alterations are evidently derived by a gene conversion mechanism involving donor sequence information contained in other V genes or pseudo V genes (Reynaud et al. 1987; Thompson and Neiman 1987; Knight and Crane 1995). Accordingly, templated mutagenesis, whether by reciprocal recombination or gene conversion, has received serious consideration as a possible mechanism for post-antigen-driven V gene diversification in mice and humans (Maizels 1995). Interest in this idea peaked following a demonstration of templated V gene mutagenesis within a murine hybridoma cell line in vitro (Krawinkel et al. 1983). More recently, two groups have demonstrated that a recombinational diversification mechanism can operate in transgenic mice that carry specialized V gene constructs (Giusti and Manser 1994; Xu and Selsing 1994). These observations suggest that somatic mutagenesis in mice and humans might include a template-directed component. However, a wide body of contrary evidence indicates that somatic diversification in these species occurs by a mechanism that more closely resembles a point mutation process. If templated mutagenesis does occur, it almost certainly involves very short template sequences and generates only a fraction of all somatic mutations.

The argument for a point-like mutation mechanism is predicated for the most part on five observations. The first of these, described earlier in this chapter, is that clusters of somatic mutations shared by independently-derived hybridomas have no genomic counterpart that might have served as a donor template (Wysocki et al. 1990). The second observation is that somatic mutations fall with a high frequency in J-C introns, which have no obvious homologous genomic counterparts that might serve as donor templates (Gearhart and Bogenhagen 1983; Hackett et al. 1990; Weber et al. 1991a, b). The third is that irrelevant sequences, such as those derived from β globin and bacterial genes, can be targeted for mutagenesis in transgenic mice by placing them within the context of Ig gene regulatory sequences that appear to direct mutagenesis (Umar et al. 1991; Azuma et al. 1993; Chun et al. 1995; Yelamos et al. 1995; Tumas-Brundage et al. 1996). The fourth comes from our recent finding that the somatic mutation mechanism exhibits a sequence-specific target bias that is the same for dissimilar sequences in light and heavy chain J-C introns and apparently for coding sequences as well (Smith et al. 1996). Finally, somatic mutations within V coding sequences often have no apparent donors within members of the same gene family (Chien et al. 1988; Ford et al. 1994b; Rogerson 1995; Vora and Manser 1995).

4.2 The Signature of Somatic Mutagenesis

A bias in the mononucleotide substitution pattern by the somatic mutation mechanism was first described by LEBECQUE and GEARHART (1990), who analyzed mutations in J-C introns, where they presumably do not often influence cellular selection and hence mutation sample. They found that the somatic mutation pattern was clearly distinguishable from the meiotic mutation pattern. Transitions were favored over transversions, and certain nucleotide substitutions were preferred, with an asymmetry that indicated a mononucleotide target bias and a strand bias. For example, A on the coding strand mutates to G more frequently than on the noncoding strand. Their findings have been repeatedly confirmed (KAARTINEN 1990; MANSER 1990; WEBER et al. 1991a, 1994; BETZ et al. 1993a; SMITH et al. 1996), and extended to human V genes (INSEL and VARADE 1994).

Efforts by two other groups to uncover sequence-specific target preferences have resulted in a tentative identification of several motifs, most notably R*G*YW (G/A G T/C A/T) and T*A*A by ROGOZIN and KOLCHANOV (1992) and AGY (A G C/T), CAGCT, and AAGTT by MILSTEIN and colleagues (BETZ et al. 1993a). The R*G*YW and T*A*A motifs were uncovered using a consensus search algorithm applied to 79 somatic mutations located within several mutated copies of a single V gene. The algorithm assumes that a mutation motif directs mutagenesis to a single nucleotide, italicized in the two cases above. When the authors then examined a larger sample of diverse V gene sequences and asked whether these motifs were more heavily mutated than expected for a random process, they found this to be the case. BETZ et al. (1993a) in contrast, used a visual approach to identify "hot spots" in coding regions of V_H and V_κ genes. They rationalized that if repeated mutations at a defined position of a putative motif were consistent with the mononucleotide substitution bias, and if the mutations revealed no evidence of selection either because they were silent, because they occurred in a codon where multiple amino acid changes were tolerated, or because they occurred in a "passenger" transgene that could not mediate positive selection of B cells, then the motif defines a hot spot of mutagenesis. Using this approach they tentatively identified the AGY, CAGCT and AAGTT motifs. In addition, they obtained some evidence that palindromic sequences direct mutagenesis to the tip of potential stem loops (BETZ et al. 1993b). The possible role of inverse and direct repeat sequences in directing mutagenesis was also suggested in studies by GOLDING et al. (1987) and KOLCHANOV et al. (1987).

In examining this work, it became apparent that the procedures used to define somatic mutation motifs were often predicated on unnecessary and risky assumptions. Using coding sequence information to define a mutation motif is risky because most base changes within codons result in amino acid replacements, and replacement somatic mutations can influence cellular selection and hence the observed mutation sample. On the other hand, excluding replacement somatic mutations from analysis to avoid exactly this problem introduces the opposite caveat that some motifs may be overlooked, a problem exacerbated by in vivo cellular selection pressures against structurally-destructive mutations or mutations that may confer autoreactive specificities, as suggested by the CASSON and MANSER (1995b) study. Neither of the approaches made a provision for identifying motifs of intermediate or low mutation frequency, primarily because of insufficient mutation sample size and qualitative limits on observable mutations due to the above-mentioned selection contingencies. Requiring that mutations at a given nucleotide position in a potential motif follow the mononucleotide target bias is a circularism, and it introduces another unnecessary restriction on observable motifs. Finally, the strategy of defining motifs in the context of a single nucleotide position is unnecessarily restrictive.

To avert these problems, SMITH et al. (1996) took a different approach to address the mutation motif issue. Selection caveats were eliminated by exclusively analyzing somatic mutations in the noncoding J-C introns of heavy and light chain V genes, where most mutations should not influence cellular selection. A large mutation sample size was analyzed, 843 somatic mutations in all, contained in over 92 000 bases of sequence. The restriction that mutations occur at a given nucleotide position of a motif was eliminated. This was achieved by recording every somatic mutation in the context of the two dinucleotides and three trinucleotides in which it fell. In essence, they determined the frequency with which every one of the 16 dinucleotides and every one of the 64 trinucleotides was mutated, regardless of mutation position. Clonal analysis of the hybridomas was performed to eliminate duplicate mutations that were apparently the consequence of a single event preceding cell division. Di- and trinucleotide mutation frequencies were normalized with respect to the frequencies with which specific di- and trinucleotides occurred in the sequences. Finally, the analysis was conducted independently in different data sets, so that hypothesized motifs derived from one data set could be tested upon analysis of other data sets. Using this approach, SMITH et al. (1996) identified unique hot mutation motifs, resolved to higher-resolution some of the proposed motifs, and excluded others. Their results indicated that di- and trinucleotide targets exhibit a continuum of mutability and that an extremely hot sequence is formed by an overlap of two hot trinucleotide targets.

Among dinucleotides, GC and TA mutated with the highest frequencies, though only about two-fold above average and about four- or five-fold above those that mutated the least (Table 1). Among trinucleotides, AGC, TAC, and their reverse complements GCT and GTA were most mutable, with frequencies that ranged from two- to three-fold above average and from four- to nine-fold higher than those of the least mutated sequences. Given the number of mutations examined, these deviations were highly significant. The TAA motif of KOLCHANOV and ROGOZIN mutated approximately one and a half times higher than average, but without the restriction that mutation be confined to the second nucleotide position. The proposed AAGTT motif occurred three times but with only two mutations, and AGT did not mutate more frequently than average, thus reducing the earlier proposed AGY motif to AGC.

An extremely high mutation frequency in the AGCT sequence was evident from the SMITH et al. (1996) analysis. This sequence is an overlap of two of the top four hottest triplets: AGC and GCT. It occurs only four times in the data set but

Table 1. Di- and trinucleotide target preferences for somatic mutagenesis [adapted from Smith et al. (1996)]

Dinucleotide observed/expected[a]		Trinucleotide observed/expected	
GC	2.05	AGC	2.68
TA	1.73	TAC	2.32
AC	1.49	GTA	2.32
AA	1.44	GCT	2.15
AT	1.33	TGC	1.90
		AAC	1.80
		CTA	1.73
		TAG	1.67
		GCA	1.62
		AAG	1.61
		ACT	1.55

This table lists some of the di- and tri-nucleotide sequences that mutate most frequently.
[a] Indicates the frequency of observed mutations divided by the frequency of mutations expected for a mutation process that has no sequence preferences. Data were obtained from 664 somatic mutations contained in 80 647 bases of sequence in hybridoma heavy and light chain J-C introns.

carries 22 somatic mutations. (In contrast, AGTT occurs seven times and contains only three somatic mutations.) We believe that AGCT accounts for much of the RGYW mutability and all of the proposed CAGCT mutability. Of the seven other sequences comprised by RGYW, four contain one of the four hottest triplets, and a fifth contains a GCA trinucleotide, which mutates approximately one and a half times above average. However, the other two sequences (GGTT, AGTT) contain no hot trinucleotides. Accordingly, the RGYW mutation motif is best reduced to its triplet composition and the overlap AGCT sequence, which accounts for most of its high mutability. Other sequences consisting of overlaps of the four hottest triplets did not occur frequently enough to draw any definite conclusion regarding their mutability. (GTAC, twice with nine mutations; GTAGC, once with five mutations; GCTAC no occurrences.)

These data have important implications regarding a mutation mechanism. Mutation target preferences were the same in both light and heavy chain J-C introns (Smith and Wysocki, unpublished data). In addition, most of the "hot-spots" described previously in V gene coding sequences (Betz et al. 1993a, b; Gonzalez-Fernandez et al. 1994) were contained in one of the four hottest triplets identified by Smith et al. (1996). These observations argue against a strict recombinational/conversion mechanism for mutagenesis, since the target sequences are divergent and the presumed template donors would have to be distinct for different V gene loci. We cannot escape this conclusion even with the provision that templated mutagenesis might occur following a preferred opening of the DNA at hot sequences. This is because the continuum of mutability for *all* di- and trinucleotides is relatively consistent between the heavy and light J-C introns, while di- and trinucleotide compositions are not (Smith and Wysocki, unpublished data). Clearly, some of the mutations must be derived in an untemplated manner, or alternatively, a universal token "template" would have to be invoked, with pairing and exchange restricted to trinucleotide or dinucleotide sequences. The data also cannot be reconciled with a model where hot sequences define sites of DNA opening, and in which the likelihood of untemplated mutagenesis is only inversely dependent upon distance from the opened site. Such a model would predict a few hot sequences, with all other sequences uniformly less mutable. This is clearly not observed.

4.3 Associations Between Somatic Mutagenesis and Ig Gene Transcription

Most early models of somatic mutagenesis invoked an error-prone process associated with DNA replication. (Brenner and Milstein 1966). However, the possibility that somatic mutagenesis might be associated with transcription has recently drawn widespread attention for two reasons. The first is the general finding that somatic mutagenesis in antibody genes appears to be correlated with Ig gene transcription. The second is the discovery that common trans-acting components play roles in transcription-coupled repair and nucleotide excision repair (Hanawalt 1994; Lehmann 1995).

Transcription appears to be necessary for mutagenesis because mutations are infrequent on the nonproductive allele when V gene assembly does not occur, and thus where the V gene promoter is not brought into proximity with enhancer elements (Gorski et al. 1983; Roes et al. 1989; Weber et al. 1991b). Somatic mutations are distributed asymmetrically, with a sharp boundary near the 5′ transcriptional initiation site and a frequency peak in the V coding region (Both et al. 1990, Lebecque and Gearhart 1990; Weber 1991b; Rothenfluh et al. 1993; Rogerson 1994). Mutation frequency gradually diminishes in the J-C intron to a distance approximately 1.5 kb 3′ of the promoter. Moving the promoter element results in a corresponding shift in the mutation frequency pattern (Weber et al. 1994; Peters et al. 1996; Tumas-Brundage et al. 1996). In mice that carry κ transgenes, the κ promoter and both identified enhancer elements appear to be necessary for full mutagenesis (Betz et al. 1994). The observed strand bias of mutagenesis suggests a possible association with transcription as well, particularly in view of the fact that transcription-coupled repair operates preferentially on the transcribed strand of DNA. In mice that carry multiple tandem copies of κ transgenes, only some of the copies are mutated in a manner that could be explained by preferential transcription due to enhancer competition (O'Brien et al. 1987; Storb et al. 1996). Finally, a transgene construct in which the natural Ig promoter was replaced with one lacking the TATA box does not sponsor a high level of mutagenesis (Tumas-Brundage et al. 1996).

While the association between somatic mutagenesis and transcription may be undeniable, it is mostly correlative. Transcription may be necessary for mutagenesis, but it is clearly not sufficient, as shown in several transgenic studies (Giusti and Manser 1993; Betz et al. 1994; Hengstschlager et al. 1994). Furthermore, the association between mutation and transcription may only be a reflection of a potential requirement for DNA accessibility, which appears to be important for V gene rearrangements and isotype switching. Finally, it is possible that mutagenesis requires transcription, if error-prone reverse transcription and integration are involved, as proposed by STEELE and POLLARD (1987).

If transcription is necessary for somatic mutation, then one might expect to find mutagenesis in centrocytes, rather than the centroblast population, as assumed by many authors, and ironically some who favor its direct association with transcription. JACOB et al. (1993) suggested that an increased frequency of hybrid PCR products seen upon V gene amplifications involving groups of centrocytes could be an indication of DNA modification, perhaps by the mutation mechanism. In ex vivo studies of single canonical mIg$^+$ cells, with phenotypic characteristics of centrocytes, we also made several puzzling observations, suggesting that these cells are possibly mutating their antibody V genes (LIU et al. 1996b). Of all sorted single canonical cells, greater than half yielded only a heavy chain PCR product or a light chain PCR product, despite the fact that they expressed both antibody chains on their surfaces, which is a necessary precondition for identification with the anti-idiotype used for sorting. In contrast, when single canonical hybridoma B cells were treated similarly, both heavy and light chain gene PCR products were obtained at a high frequency (85%). In most cases a hybridoma cell that failed to yield one PCR product, also failed to yield the other, indicating that the cell was not successfully transferred from slide to microtube with lysis buffer (LIU et al. 1992b; JENA et al. 1996). The ex vivo isolated cells were visually large, expressed Ig transcripts, and so did not appear to be apoptotic. The possibility that unamplifiable V genes in these cells were modified by a mutageneic complex is one interesting interpretation that is consistent with these observations.

We also observed evidence for κ transcript heterogeneity in a single canonical B cell, from which single transcript molecules were amplified by RT-PCR from highly-diluted aliquots of cytoplasmic extract (JENA and WYSOCKI, unpublished data). By a Poisson analysis, the cell contained several hundred cytoplasmic κ transcripts. Sequencing of the amplified κ gene revealed one somatic mutation that was shared by all of the 37 sequenced V κ transcripts, which were also directly sequenced without cloning. However, several of the κ PCR products revealed point mutations and rearrangements. Some of the rearrangements appeared to be due to activation of cryptic splice sites, since they followed the GT/AG splicing rule. The frequency of point mutations was approximately five times higher than expected from reverse transcription and Taq amplification, and none of them were the most common type of Taq misincorporations (T to C or G to A). Similar sequencing of single amplified β2-microglobulin cDNA molecules revealed only one type of nucleotide change, apparently due to "template sliding", at a stretch of seven consecutive A nucleotides. Moreover, we have found that Taq misincorporation errors that arise when amplifying single strands of cDNA often produce dual signals at one mobility location on the sequencing autoradiogram. Only one such presumed Taq polymerase error was seen in 11 000 bases of V_{κ} sequence, which fits well with the error rate for this enzyme, calculated by others and observed by us in preceding studies.

Since none of the heterogeneous transcript mutations were found in the κ gene, the above observations suggest that they were either directly introduced in the RNA at a high rate, possibly by RNA polymerase, or that the transcribed strand of the V_κ gene was being mutated and repaired at a high rate. It is noteworthy that some of the mIg^+ canonical cells carried hundreds or thousands of κ transcripts, despite no apparent antibody secretion and a very low level of surface idiotype expression, consistent with the general observation that centrocytes express only low levels of immunoglobulin. It is puzzling why cells with so little antibody production carry so many antibody gene transcripts. If transcript abundance in these cells is a reflection of transcriptional activity, this observation might suggest that Ig gene transcription is playing a cryptic role, perhaps in V gene mutagenesis. It is an attractive thought, given the aforementioned indirect links between transcription and mutagenesis.

5 Conclusions and Prospects

Memory B lymphocytes of mice and humans are carved out of a diversified repertoire by a somatic evolutionary process. The selection pressures that shape the memory pool include an intense component for improvement in receptor affinity, and probably a counter-selecting pressure against autoreactivity, though decisive evidence for the latter is lacking. In addition, since somatic mutagenesis creates potential V region epitopes for T helper cell cognition, it is possible that the B cell may face yet a second counter-selection pressure against such epitopes, but this is purely conjecture.

The population dynamics that underlie B cell selection during memory development have been significantly advanced by molecular studies of GC B cells and single sorted memory progenitors. It is clear that GCs are sites of somatic mutagenesis, development, and selection of memory B cells, but whether they are exclusive sites for these cardinal events is presently unclear. A major point of concern is the paradox that while somatic mutagenesis and selection appear to occur over multiple division cycles, mIg^+ centrocytes in the GC appear to divide little if at all, while GC centroblasts divide rapidly but seem to lack mIg. Cyclic re-entry of positively selected centrocytes to the centroblast compartment could explain this paradox, and accordingly, testing this hypothesis is critical to understanding the population dynamics behind memory cell generation. An important component of the cyclic-re-entry hypothesis is that Ig synthesis is alternately induced and repressed in GC B cells. Recent studies of Ig transcript processing in transgenic models have revealed that B cells can indeed regulate Ig transcript abundance by novel mechanisms. Cycles of Ig transcript induction and purging of old transcripts between rounds of mutagenesis could be a mechanism that ensures efficient stepwise selection of B cells. Thus, it is of primary importance to examine Ig transcript regulation by GC B cells in the context of the cyclic re-entry idea.

The nature of the somatic mutation mechanism remains an enigma to the discipline of immunology and to biology in a larger sense, since there is no established precedent for a physiological mutator. The ability to direct mutations to non-Ig transgenes and a consistent hierarchy of mutability among di- and trinucleotides in divergent Ig sequences argues strongly for a point mutation mechanism. This must be qualified, however, since available data do not formally exclude the possibility that a subset of "somatic mutations" are derived by a templated recombinational process. Several striking associations between transcription and mutagenesis of Ig genes have led to the hypothesis that transcription may be directly involved in somatic mutagenesis. In part, this hypothesis is fueled by recent discoveries linking transcription-coupled repair to nucleotide excision repair in other cells and organisms. If correct, the mIg^+ centrocyte could be the mutating cell, rather than the commonly assumed centroblast (mIg^-). On the other hand, it is possible that the associations between mutation and transcription are indirect. For example, transcription may play a role only in allowing mutator accessibility to Ig genes. Recent advances in mouse genome manipulation and the identification of cell lines that mutate their antibody genes at a high rate (Green et al. 1995; Zhu et al. 1996) should provide avenues to resolve this issue and to finally elucidate the mechanism of antibody V gene somatic mutagenesis.

Acknowledgements. Supported by grants from the NIH R01AI39563, 2R01AI33613 and NIH PO1AI 22295

References

Allen D, Simon T, Sablitzky F, Rajewsky K, Cumano A (1988) Antibody engineering for the analysis of affinity maturation of an anti-hapten response. EMBO J 7:1995–2001

Aoufouchi S, Yelamos J, Milstein C (1996) Nonsense mutations inhibit RNA splicing in a cell-free system: recognition of mutant codon is independent of protein synthesis. Cell 85:415–422

Azuma T, Motoyama N, Fields LE, Loh DY (1993) Mutations of the chloramphenicol acetyl transferase transgene driven by the immunoglobulin promoter and intron enhancer. Int Immunol 5:121–130

Berek C, Milstein C (1987) Mutation drift and repertoire shift in the maturation of the immune response. Immunol Rev 96:23–41

Berek C, Ziegner M (1993) The maturation of the immune response. Immunol Today 14:400–404

Betz AG, Neuberger MS, Milstein C (1993a) Discriminating intrinsic and antigen-selected mutational hotspots in immunoglobulin V genes. Immunol Today 14:405–411

Betz AG, Rada C, Pannell R, Milstein C, Neuberger MS (1993b) Passenger transgenes reveal intrinsic specificity of the antibody hypermutation mechanism: clustering, polarity, and specific hot spots. Proc Natl Acad Sci USA 90:2385–2388

Betz AG, Milstein C, Gonzalez-Fernandez A, Pannell R, Larson T, Neuberger MS (1994) Elements regulating somatic hypermutation of an immunoglobulin kappa gene: critical role for the intron enhancer/matrix attachment region. Cell 77:239–248

Blier PR, Bothwell A (1987) A limited number of B cell lineages generates the heterogeneity of a secondary immune response. J Immunol 139:3996–4006

Both GW, Taylor L, Pollard JW, Steele EJ (1990) Distribution of mutations around rearranged heavy-chain antibody variable-region genes. Mol Cell Biol 10:5187–5196

Brenner S, Milstein C (1966) Origin of antibody variation. Nature 211:242–243

Butcher EC, Rouse RV, Coffman RL, Nottenburg CN, Hardy RR, Weissman IL (1982) Surface phenotype of Peyer's patch germinal center cells: implications for the role of germinal centers in B cell differentiation. J Immunol 129:2698–2707
Casson LP, Manser T (1995a) Evaluation of loss and change of specificity resulting from random mutagenesis of an antibody VH region. J Immunol 155:5647–5654
Casson LP, Manser T (1995b) Random mutagenesis of two complementarity determining region amino acids yields an unexpectedly high frequency of antibodies with increased affinity for both cognate antigen and autoantigen. J Exp Med 182:743–750
Chen C, Roberts VA, Rittenberg MB (1992) Generation and analysis of random point mutations in an antibody CDR2 sequence: many mutated antibodies lose their ability to bind antigen. J Exp Med 176:855–866
Chicz RM, Urban RG, Lane WS et al (1992) Predominant naturally processed peptides bound to HLA-DR1 are derived from MHC-related molecules and are heterogeneous in size. Nature 358:764–768
Chien NC, Pollock RR, Desaymard C, Scharff MD (1988) Point mutations cause the somatic diversification of IgM and IgG2a antiphosphorylcholine antibodies. J Exp Med 167:954–973
Chun EM, Zwollo P, Wong P, Love C, Doft A, Pollock RR (1995) Analysis of somatic mutation with a reporter gene construct. Ann N Y Acad Sci 764:183
Claflin JL, Berry J, Flaherty D, Dunnick W (1987) Somatic evolution of diversity among antiphosphocholine antibodies induced with Proteus morganii. J Immunol 138:3060–3068
Clarke SH, Huppi K, Ruezinsky D, Staudt L, Gerhard W, Weigert M (1985) Inter- and intraclonal diversity in the antibody response to influenza hemagglutinin. J Exp Med 161:687–704
Crews S, Griffin J, Huang H, Calame K, Hood L (1981) A single VH gene segment encodes the immune response to phosphorylcholine: somatic mutation is correlated with the class of the antibody. Cell 25:59–66
Desiderio SV, Yancopoulos GD, Paskind M, Thomas E, Boss MA, Landau N, Alt FW, Baltimore D (1984) Insertion of N regions into heavy-chain genes is correlated with expression of terminal deoxytransferase in B cells. Nature 311:752–755
Diamond B, Scharff MD (1984) Somatic mutation of the T15 heavy chain gives rise to an antibody with autoantibody specificity. Proc Natl Acad Sci USA 81:5841–5844
Durdik J, Gerstein RM, Rath S, Robbins PF, Nisonoff A, Selsing E (1989) Isotype switching by a microinjected μ immunoglobulin heavy chain gene in transgenic mice. Proc Natl Acad Sci USA 86:2346–2350
Ellenberger J, Creadon G, Zhang X, Wysocki LJ (1993) Recruiting memory B cells with changed antigenic specificity. J Immunol 151:5272–5281
Eyerman MC, Zhang X, Wysocki LJ (1996) T cell recognition and tolerance of antibody diversity. J Immunol 157:1037–1046
Fish S, Zenowich E, Fleming M, Manser T (1989) Molecular analysis of original antigenic sin. I. Clonal selection, somatic mutation, and isotype switching during a memory B cell response. J Exp Med 170:1191–1209
Fish S, Fleming M, Sharon J, Manser T (1991) Different epitope structures select distinct mutant forms of an antibody variable region for expression during the immune response. J Exp Med 173:665–672
Fliedner TM, Kress M, Cronkite EP, Robertons JS (1964) Cell proliferation in germinal centers of the rat spleen Ann N Y Acad Sci 113:578–594
Ford JE, McHeyzer-Williams MG, Lieber MR (1994a) Chimeric molecules created by gene amplification interfere with the analysis of somatic hypermutation of murine immunoglobulin genes. Gene 142:279–283
Ford JE, McHeyzer-Williams MG, Lieber MR (1994b) Analysis of individual immunoglobulin λ light chain genes amplified from single cells is inconsistent with variable region gene conversion in germinal-center B cell somatic mutation. Eur J Immunol 24:1816–1822
French DL, Laskov R, Scharff MD (1989) The role of somatic hypermutation in the generation of antibody diversity. Science 244:1152–1157
Gearhart PJ, Bogenhagen DF (1983) Clusters of point mutations are found exclusively around rearranged antibody variable genes. Proc Natl Acad Sci USA 80:3439–3443
Gerstein RM, Frankel WN, Hsieh CL, Durdik JM, Rath S, Coffin JM, Nisonoff A, Selsing E (1990) Isotype switching of an immunoglobulin heavy chain transgene occurs by DNA recombination between different chromosomes. Cell 63:537–548
Gilbert W (1978) Why are genes in pieces? Nature 271:501

Giusti AM, Manser T (1993) Hypermutation is observed only in antibody H chain V region transgenes that have recombined with endogenous immunoglobulin H DNA: implications for the location of cis-acting elements required for somatic mutation. J Exp Med 177:797–809
Giusti AM, Manser T (1994) Somatic generation of hybrid antibody H chain genes in transgenic mice via interchromosomal gene conversion. J Exp Med 179:235–248
Golding GB, Gearhart PJ, Glickman BW (1987) Patterns of somatic mutations in immunoglobulin variable genes. Genetics 115:169–176
Gonzalez-Fernandez A, Gupta SK, Pannell R, Neuberger MS, Milstein C (1994) Somatic mutation of immunoglobulin λ chains: a segment of the major intron hypermutates as much as the complementarity-determining regions. Proc Natl Acad Sci USA 91:12614–12618
Goodnow CC (1996) Balancing immunity and tolerance: deleting and tuning lymphocyte repertoires. Proc Natl Acad Sci USA 93:2264–2271
Gorski J, Rollini P, Mach B (1983) Somatic mutations of immunoglobulin variable genes are restricted to the rearranged V gene. Science 220:1179–1181
Green NS, Rabinowitz JL, Zhu M, Kobrin BJ, Scharff MD (1995) Immunoglobulin variable region hypermutation in hybrids derived from a pre-B- and a myeloma cell line. Proc Natl Acad Sci USA 92:6304–6308
Griffiths GM, Berek C, Kaartinen M, Milstein C (1984) Somatic mutation and the maturation of immune response to 2-phenyl oxazolone. Nature 312:271–275
Hackett J Jr, Rogerson BJ, O'Brien RL, Storb U (1990) Analysis of somatic mutations in κ transgenes. J Exp Med 172:131–137
Han S, Hathcock K, Zheng B, Kepler TB, Hodes R, Kelsoe G (1995a) Cellular interaction in germinal centers. Roles of CD40 ligand and B7-2 in established germinal centers. J Immunol 155:556–567
Han S, Zheng B, Dal Porto J, Kelsoe G (1995b) In situ studies of the primary immune response to (4-hydroxy-3-nitrophenyl)acetyl. IV. Affinity-dependent, antigen-driven B cell apoptosis in germinal centers as a mechanism for maintaining self-tolerance. J Exp Med 182:1635–1644
Hanawalt PC (1994) Transcription-coupled repair and human disease. Science 266:1957–1958
Hengstschlager M, Williams M, Maizels N (1994) A λ 1 transgene under the control of a heavy chain promoter and enhancer does not undergo somatic hypermutation. Eur J Immunol 24:1649–1656
Insel RA, Varade WS (1994) Bias in somatic hypermutation of human VH genes. Int Immunol 6:1437–1443
Jacob J, Kelsoe G (1992) In situ studies of the primary immune response to (4-hydroxy-3-nitrophenyl)acetyl. II. A common clonal origin for periarteriolar lymphoid sheath-associated foci and germinal centers. J Exp Med 176:679–687
Jacob J, Kassir R, Kelsoe G (1991) In situ studies of the primary immune response to (4-hydroxy-3-nitrophenyl)acetyl. I. The architecture and dynamics of responding cell populations. J Exp Med 173:1165–1175
Jacob J, Przylepa J, Miller C, Kelsoe G (1993) In situ studies of the primary immune response to (4-hydroxy-3-nitrophenyl)acetyl. III. The kinetics of V region mutation and selection in germinal center B cells. J Exp Med 178:1293–1307
Jena PK, Liu AH, Smith DS, Wysocki LJ (1996) Amplification of genes, single transcripts and cDNA libraries from one cell and direct sequence analysis of amplified products derived from one molecule. J Immunol Methods 190:199–213
Jorgensen T, Bogen B, Hannestad K (1983) T helper cells recognize an idiotope located on peptide 88-114/117 of the light chain variable domain of an isologous myeloma protein (315). J Exp Med 158:2183–2188
Kaartinen M (1990) Characteristics of selection-free mutations and effects of subsequent selection. In: Steele EJ (ed) Somatic hypermutation in V genes. CRC Press, Boca Raton, pp141–162
Kelsoe G (1995) In situ studies of the germinal center reaction. Adv Immunol 60:267–288
Kelsoe G (1996a) Life and death in germinal centers (redux). Immunity 4:107–111
Kelsoe G (1996b) The germinal center: a crucible for lymphocyte selection. Semin Immunol 8:179–184
Kepler TB, Perelson AS (1993) Cyclic re-entry of germinal center B cells and the efficiency of affinity maturation. Immunol Today 14:412–415
Klinman NR (1996) The "clonal selection hypothesis" and current concepts of B cell tolerance. Immunity 5:189–195
Knight KL, Crane MA (1995) Development of the antibody repertoire in rabbits. Ann N Y Acad Sci 764:198–206
Kolchanov NA, Solovyov VV, Rogozin IB (1987) Peculiarities of immunoglobulin gene structures as a basis for somatic mutation emergence. FEBS Lett 214:87–91

Krawinkel U, Zoebelein G, Bruggemann M, Radbruch A, Rajewsky K (1983) Recombination between antibody heavy chain variable-region genes: evidence for gene conversion. Proc Natl Acad Sci USA 80:4997–5001
Kussie PH, Parhami-Seren B, Wysocki LJ, Margolies MN (1994) A single engineered amino acid substitution changes antibody fine specificity. J Immunol 152:146–152
Lebecque SG, Gearhart PJ (1990) Boundaries of somatic mutation in rearranged immunoglobulin genes: 5′ boundary is near the promoter, and 3′ boundary is approximately 1 kb from V(D)J gene. J Exp Med 172:1717–1727
Lehmann AR (1995) Nucleotide excision repair and the link with transcription. Trends Biochem Sci 20:402–405
Lieber M (1996) Immunoglobulin diversity: rearranging by cutting and repairing. Curr Biol 6:134–136
Linton PJ, Rudie A, Klinman NR (1991) Tolerance susceptibility of newly generating memory B cells. J Immunol 146:4099–4104
Liu YJ, Zhang J, Lane PJ, Chan EY, MacLennan IC (1991) Sites of specific B cell activation in primary and secondary responses to T cell-dependent and T cell-independent antigens. Eur J Immunol 21:2951–2962
Liu YJ, Johnson GD, Gordon J, MacLennan IC (1992a) Germinal centres in T-cell-dependent antibody responses. Immunol Today 13:17–21
Liu AH, Creadon G, Wysocki LJ (1992b) Sequencing heavy- and light-chain variable genes of single B-hybridoma cells by total enzymatic amplification. Proc Natl Acad Sci USA 89:7610–7614
Liu YJ, Malisan F, de Bouteiller O, Guret C, Lebecque S, Banchereau J, Mills FC, Max EE, Martinez-Valdez H (1996a) Within germinal centers, isotype switching of immunoglobulin genes occurs after the onset of somatic mutation. Immunity 4:241–250
Liu AH, Jena PK, Wysocki LJ (1996b) Tracing the development of single memory-lineage B cells in a highly defined immune response. J Exp Med 183:2053–2063
Lozano F, Rada C, Jarvis JM, Milstein C (1993) Affinity maturation leads to differential expression of multiple copies of a kappa light-chain transgene. Nature 363:271–273
MacLennan IC (1994) Germinal centers. Annu Rev Immunol 12:117–39
Maizels N (1995) Somatic hypermutation: how many mechanisms diversify V region sequences? Cell 83:9–12
Malipiero UV, Levy NS, Gearhart PJ (1987) Somatic mutation in anti-phosphorylcholine antibodies. Immunol Rev 96:59–74
Manser T (1989) Evolution of antibody structure during the immune response. The differentiative potential of a single B lymphocyte. J Exp Med 170:1211–1230
Manser T (1990) Regulation, timing and mechanism of antibody V gene somatic hypermutation: lessons from the arsonate system. In: Steele EJ (ed) Somatic hypermutation in V genes. CRC Press, Boca Raton, pp41–54
Manser T, Wysocki LJ, Margolies MN, Gefter ML (1987) Evolution of antibody variable region structure during the immune response. Immunol Rev 96:141–162
Marrack P, Ignatowicz L, Kappler JW, Boymel J, Freed JH (1993) Comparison of peptides bound to spleen and thymus class II. J Exp Med 178:2173–2183
Matsumoto M, Lo SF, Carruthers CJ, Min J, Mariathasan S, Huang G, Plas DR, Martin SM, Geha RS, Nahm MN, Chaplin DD (1996) Affinity maturation without germinal centres in lymphotoxin-α - deficient mice. Nature 382:462–466
McHeyzer-Williams MG, Nossal GJ, Lalor PA (1991) Molecular characterization of single memory B cells. Nature 350:502–505
McHeyzer-Williams MG, McLean MJ, Lalor PA, Nossal GJ (1993) Antigen-driven B cell differentiation in vivo. J Exp Med 178:295–307
McKean D, Huppi K, Bell M, Staudt L, Gerhard W, Weigert M (1984) Generation of antibody diversity in the immune response of BALB/c mice to influenza virus hemagglutinin. Proc Natl Acad Sci USA 81:3180–3184
Nemazee D (1993) Promotion and prevention of autoimmunity by B lymphocytes. Curr Opin Immunol 5:866–872
Nieuwenhuis P, Opstelten D (1984) Functional anatomy of germinal centers. Am J Anat 170:421–435
Nossal GJ (1994) Negative selection of lymphocytes. Cell 76:229–239
Nossal GJ, Karvelas M, Pulendran B (1993) Soluble antigen profoundly reduces memory B-cell numbers even when given after challenge immunization. Proc Natl Acad Sci USA 90:3088–3092
O'Brien RL, Brinster RL, Storb U (1987) Somatic hypermutation of an immunoglobulin transgene in kappa transgenic mice. Nature 326:405–409

Parhami-Seren B, Wysocki LJ, Margolies MN, Sharon J (1990) Clustered H chain somatic mutations shared by anti-p-azophenylarsonate antibodies confer enhanced affinity and ablate the cross-reactive idiotype. J Immunol 145:2340–2346
Pascual V, Liu YJ, Magalski A, de Bouteiller O, Banchereau J, Capra JD (1994) Analysis of somatic mutation in five B cell subsets of human tonsil. J Exp Med 180:329–339
Peters A, Storb U (1996) Somatic hypermutation of immunoglobulin genes is linked to transcription initiation. Immunity 4:57–65
Pulendran B, Kannourakis G, Nouri S, Smith KG, Nossal GJ (1995) Soluble antigen can cause enhanced apoptosis of germinal-centre B cells. Nature 375:331–334
Rajewsky K (1996) Clonal selection and learning in the antibody system. Nature 381:751–758
Reynaud CA, Anquez V, Grimal H, Weill JC (1987) A hyperconversion mechanism generates the chicken light chain preimmune repertoire. Cell 48:379–388
Reynaud CA, Garcia C, Hein WR, Weill JC (1995) Hypermutation generating the sheep immunoglobulin repertoire is an antigen-independent process. Cell 80:115–125
Rickert RC, Wloch MK, Hahn RL, Clarke SH (1995) Binding analysis of antibodies produced by precursor and branchpoint intermediates of an anti-influenza hemagglutinin B cell clone. Parallel replacement mutations do not confer increased avidity for hemagglutinin. J Immunol 154:2209–2216
Roes J, Huppi K, Rajewsky K, Sablitzky F (1989) V gene rearrangement is required to fully activate the hypermutation mechanism in B cells. J Immunol 142:1022–1026
Rogerson BJ (1994) Mapping the upstream boundary of somatic mutations in rearranged immunoglobulin transgenes and endogenous genes. Mol Immunol 31:83–98
Rogerson BJ (1995) Somatic hypermutation of VHS107 genes is not associated with gene conversion among family members. Int Immunol 7:1225–1235
Rogozin IB, Kolchanov NA (1992) Somatic hypermutagenesis in immunoglobulin genes. II. Influence of neighbouring base sequences on mutagenesis. Biochim Biophys Acta 1171:11–18
Rose ML, Birbeck MS, Wallis VJ, Forrester JA, Davies AJ (1980) Peanut lectin binding properties of germinal centres of mouse lymphoid tissue. Nature 284:364–366
Rothenfluh HS, Taylor L, Bothwell AL, Both GW, Steele EJ (1993) Somatic hypermutation in 5′ flanking regions of heavy chain antibody variable regions. Eur J Immunol 23:2152–2159
Rudensky AYu, Preston Hurlburt P, Hong SC, Barlow SC, Janeway CA, Jr. (1991) Sequence analysis of peptides bound to MHC class II molecules. Nature 353:622–627
Schroder AE, Greiner A, Seyfert C, Berek C (1996) Differentiation of B cells in the nonlymphoid tissue of the synovial membrane of patients with rheumatoid arthritis. Proc Natl Acad Sci USA 93:221–225
Sharon J, Gefter ML, Wysocki LJ, Margolies MN (1989) Recurrent somatic mutations in mouse antibodies to p-azophenylarsonate increase affinity for hapten. J Immunol 142:596–601
Shokat KM, Goodnow CC (1995) Antigen-induced B-cell death and elimination during germinal-centre immune responses. Nature 375:334–338
Slaughter CA, Capra JD (1983) Amino acid sequence diversity within the family of antibodies bearing the major antiarsonate cross-reactive idiotype of the A strain mouse. J Exp Med 158:1615–1634
Smith DS, Creadon G, Jena PK, Portanova JP, Kotzin BL, Wysocki LJ (1996) Di- and trinucleotide target preferences of somatic mutagenesis in normal and autoreactive B cells. J Immunol 156:2642–2652
Steele EJ, Pollard JW (1987) Hypothesis: somatic hypermutation by gene conversion via the error prone DNA–RNA–DNA information loop. Mol Immunol 24:667–673
Storb U, Peters A, Klotz E, Rogerson B, Hackett J Jr. (1996) The mechanism of somatic hypermutation studied with transgenic and transfected target genes. Semin Immunol 8:131–140
Thompson CB, Neiman PE (1987) Somatic diversification of the chicken immunoglobulin light chain gene is limited to the rearranged variable gene segment. Cell 48:369–378
Tonegawa S (1983) Somatic generation of antibody diversity. Nature 302:575–581
Tumas-Brundage K, Vora KA, Giusti AM, Manser T (1996) Characterization of the cis-acting elements required for somatic hypermutation of murine antibody V genes using conventional transgenic and transgene homologous recombination approaches. Semin Immunol 8:141–150
Umar A, Schweitzer PA, Levy NS, Gearhart JD, Gearhart PJ (1991) Mutation in a reporter gene depends on proximity to and transcription of immunoglobulin variable transgenes. Proc Natl Acad Sci USA 88:4902–4906
Vora KA, Manser T (1995) Altering the antibody repertoire via transgene homologous recombination: evidence for global and clone-autonomous regulation of antigen-driven B cell differentiation. J Exp Med 181:271–281

Weber JS, Berry J, Manser T, Claflin JL (1991a) Position of the rearranged V kappa and its 5′ flanking sequences determines the location of somatic mutations in the J kappa locus. J Immunol 146:3652–3655

Weber JS, Berry J, Litwin S, Claflin JL (1991b) Somatic hypermutation of the JC intron is markedly reduced in unrearranged kappa and H alleles and is unevenly distributed in rearranged alleles. J Immunol 146:3218–3226

Weber JS, Berry J, Manser T, Claflin JL (1994) Mutations in Ig V(D)J genes are distributed asymmetrically and independently of the position of V(D)J. J Immunol 153:3594–3602

Weigert MG, Cesari IM, Yonkovich SJ, Cohn M (1970) Variability in the λ light chain sequences of mouse antibody. Nature 228:1045–1047

Weiss S, Bogen B (1991) MHC class II-restricted presentation of intracellular antigen. Cell 64:767–776

Wilson M, Hsu E, Marcuz A, Courtet M, Du Pasquier L, Steinberg C (1992) What limits affinity maturation of antibodies in Xenopus – the rate of somatic mutation or the ability to select mutants? EMBO J 11:4337–4347

Wysocki L, Manser T, Gefter ML (1986) Somatic evolution of variable region structures during an immune response. Proc Natl Acad Sci USA 83:1847–1851

Wysocki LJ, Gridley T, Huang S, Grandea AGd, Gefter ML (1987) Single germline VH and V kappa genes encode predominating antibody variable regions elicited in strain A mice by immunization with p-azophenylarsonate. J Exp Med 166:1–11

Wysocki LJ, Gefter ML, Margolies MN (1990) Parallel evolution of antibody variable regions by somatic processes: consecutive shared somatic alterations in VH genes expressed by independently generated hybridomas apparently acquired by point mutation and selection rather than by gene conversion. J Exp Med 172:315–323

Xu B, Selsing E (1994) Analysis of sequence transfers resembling gene conversion in a mouse antibody transgene. Science 265:1590–1593

Yelamos J, Klix N, Goyenechea B et al. (1995) Targeting of non-Ig sequences in place of the V segment by somatic hypermutation. Nature 376:225–229

Zhang J, MacLennan IC, Liu YJ, Lane PJ (1988) Is rapid proliferation in B centroblasts linked to somatic mutation in memory B cell clones? Immunol Lett 18:297–299

Zhu M, Green NS, Rabinowitz JL, Scharff MD (1996) Differential V region mutation of two transfected Ig genes and their interaction in cultured B cell lines. EMBO J 15:2738–2747

Repertoire Diversification of Primary vs Memory B Cell Subsets

N.R. KLINMAN

1 Introduction

The immune system accomplishes the specific recognition of foreign antigens, in the absence of reactivity to self antigens, by: (a) creating an enormous repertoire of cells each expressing an unique variable (V) region (clonotype); (b) purging the repertoire of cells whose V regions recognize self-antigenic determinants, and (c) enabling each immunogen to selectively stimulate only those cells whose V regions are high affinity for determinants of that immunogen. While this overall strategy is pervasive among both the B and T cell systems of immunocompetent animals, there is enormous variation in the extent of repertoire diversity and the means by which repertoire diversity is achieved, even among the B cell subsets of an individual. Thus, prior to any overt antigenic stimulation, the murine B cell system consists of at least four distinct B cell subsets differing in repertoire diversity and responsiveness to antigenic stimulation. After immunization, the repertoire is supplemented by the generation of memory B cells, which in some, but not all cases, enables the refinement of that portion of the repertoire that initially recognized the immunogen. The existence of a mechanism that improves the pre-existing repertoire suggests that the initial repertoire may have been functionally deficient in the spectrum of V regions capable of high affinity recognition of certain antigens. In

Department of Immunology, The Scripps Research Institute, La Jolla, CA 92037, USA this sense, since the generation of memory B cells can fill these gaps, the strategy for V region diversification used to generate and propagate memory B cells would appear to be both unique and extremely powerful. In order to place both the need and mechanism for memory B cell repertoire generation into context, it is necessary not only to understand memory B cell generation, but also the establishment and extent of diversity in the subsets of naive B cells that pre-exist antigenic stimulation.

2 Repertoire Expression in Primary B Cell Subsets

2.1 Repertoire Diversification in Conventional AFC Precursors

Conventional bone marrow-derived primary antibody forming cell (AFC) precursors represent the predominant B cell subset of adult mice. The mechanism by which this B cell subset is generated in the adult bone marrow ensures maximum repertoire diversity so that the number of V region clonotypes approaches the number of total B cells (2×10^8) or one cell per clonotype (CANCRO et al. 1978; SIGAL and KLINMAN 1978, EARLY et al. 1980; TONEGAWA 1983; ALT et al.1984; DECKER et al. 1991a). This is accomplished primarily by the process of V(D)J recombination of heavy (H) and light (L) chain V region gene segments (CREWS et al. 1981; KUROSAWA and TONEGAWA 1982; TONEGAWA 1983; ALT et al.1984; MAX 1984) which creates sufficient diversity to ensure that any given clonotype should not recur. A second means by which this is accomplished is the restricting of cell division (clonal expansion) to the pro- and pre-B cell stages of B cell development (PARK and OSMOND 1989; OSMOND 1991; DECKER et al. 1991a).

Because of this, while numerous cells may share H chains, there is no clonal expansion of cells once they have independently expressed L chains so that no two cells will have precisely the same H as well as L chains (CATON 1990; DECKER et al. 1991a). Because of the availability of 15 D_H gene segments (ICHIHARA et al. 1989; FEENEY and RIBLET 1993) and four J_H (TONEGAWA 1983), junctional imprecision, and N additions (DESIDERIO et al. 1984), the number of distinct H chain third complimentarity determining regions (HCDR3), created by the rearrangement of any of the 10^2–10^3 V_H gene segments to D_H-J_H, exceeds 10^5 (DECKER et al. 1991a). Thus, in the mouse, the number of distinct H chain V regions is likely to exceed 10^8. Since V_L–J_L recombinations is also likely to create more than 10^3–10^4 distinct L chains (MAX 1984), the total potential for Ig V region diversity is likely to exceed 10^{12}. Because of this, in the adult marrow recurrence of the same V region within the lifetime of a given animal or among members of an inbred strain should be a very rare event. However, since V gene segment rearrangement is not a random process (ALT et al. 1984; PERLMUTTER et al. 1985; DECKER et al. 1991a; MILSTEIN et al. 1992) and cells are highly selected for maturational progression on the basis of their expressed H chains (MALYNN et al. 1990; DECKER et al. 1991a; DECKER et al. 1991b, 1995b; HUETZ et al. 1993), the ideal of a maximally diverse and totally non-recurrent clonotype repertoire is probably never achieved. Indeed, in certain instances, especially when sequences do not include N additions, H chain V regions and Ig clonotypes do reproducibly recur (Blomberg et al. 1972; Pawlak and Nisonoff 1973; Lieberman et al. 1974; Makela and Karajalainen 1977; Klinman and Stone 1983; Perlmutter et al. 1984; Froscher and Klinman 1985; Clarke and McCray 1993). Such clonotype recurrence may be highly important biologically and applies mainly to the neonate and the B-1 subset of adults (see below), but can also be observed in newly generating bone marrow cells of adults as well (Klinman and Stone 1983; Froscher and Klinman 1985; Clarke and McCray 1993).

In addition to the expression of recurrent clonotypes, which presumably are present in only a few percent of adult conventional B cells, repertoire diversity of conventional B cells is markedly skewed by the preferential utilization of certain V_H, D_H and J_H segments and V_L and J_L segments as well (Decker et al. 1991a; Milstein et al. 1992). In some instances, this preferential utilization may correlate with the position of gene segments on the chromosome (Alt et al. 1984; Perlmutter et al. 1985). However, other mechanisms that impact gene accessibility must also play a role. For example, in analyses of the frequency with which V_H gene segments occur among non-productive V_H-D_H-J_H rearrangements V_H81X, the most D_H proximal V_H gene segment is overutilized (Alt et al. 1984; Decker et al. 1991a, b; Huetz et al. 1993). However, among V_H36–60 V_H gene segments V_H1210.7 is more frequently utilized, especially in neonates than more D_H proximal members of the same family (Riley et al. 1986; Decker et al. 1991a). In the case of L chains and H chain D and J segments, utilization may also be skewed by gene replacement reactions that occur subsequent to the first D_H-J_H or V_L-J_L rearrangement (Reth et al. 1986; Fitts and Mage 1995). It should be noted that (V_L-J_L) replacement reactions (receptor editing) may be induced by contact of sIg receptors of immature B cells with self-antigen (Radic et al. 1993; Tiegs et al. 1993).

The primary conventional B cell repertoire is also greatly impacted by selective events that occur after H chain expression and then again after sIg expression. It is now clear that pre-B cells vary in their ability to mature based on the sequence of their nascent H chain (Decker et al. 1991a, b; Clarke and McCray 1993; Huetz et al. 1993). In certain instances the failure to mature appears to be the result of failure of the H chain to associate with a surrogate L chain (SL) (the product of V pre-B and λ5) (Kenya et al. 1995; Klinman et al. 1997) since an H chain SL complex facilitates clonal maturation (Kitamura et al. 1992; Rolink et al. 1993). However, failure of H chains to mediate clonal maturation may be due to reasons other than failure to assemble with SL (Clarke and McCray 1993). Since among H chains that fail to mediate clonal maturation are most of those encoded by V_H81X (which represents up to 20% of all rearranged V_H gene segments), it is likely that a large percentage of nascent H chains are "dysfunctional" in mediating clonal maturation (Decker et al. 1991b; Huetz et al. 1993). In particular, it appears that the ability to facilitate clonal maturation is dependent on an appropriate HCDR3 sequence (Decker et al. 1991a, b; Clarke and McCray 1993). It is possible, therefore, that because of the extraordinary variability built into the encoding of HCDR3, many resultant H chains would fail to fold properly. In this sense, the ability to interact with SL may act as a quality control for newly generated H chains and, in particular, assess the potential of H chains to ultimately assemble with L chains. By negatively selecting cells with "dysfunctional" H chains prior to clonal expansion, H chains that enter the pool of fully expanded pre-B cell clones can be pre-selected for their ultimate potential to encode functional antibodies (Decker et al. 1991a, b).

The conventional naive B cell repertoire undergoes further pruning after L chain rearrangement and sIg expression via the negative selection of cells expressing clonotypes specific for self antigens. The mechanism, extent, and consequences of tolerance induction of immature B cells has been extensively reviewed elsewhere (Metcalf et al. 1978; Nossal 1983; Nemazee and Burki 1989; Goodnow 1992; Hartley et al. 1993; Klinman 1996a). Notwithstanding results from some sIg transgenic mice (Russell et al. 1991), the elimination of newly generating B cells appears to be both highly specific and affinity dependent (Metcalf and Klinman 1976, 1977; Teale and Klinman 1980, 1984; Riley and Klinman 1986). This conclusion is consistent with the finding that B cells with low affinity for self antigens can be easily demonstrated in the peripheral B cell pool (Harris et al. 1982; Stockinger and Hausmann 1988; Cooper et al. 1988). This has led to the hypothesis that permissiveness for cells with low-affinity for the innumerable spectrum of self antigenic determinants ensures the expression of a diverse (but selected) repertoire.

A final level of selection of the repertoire of mature conventional B cells concerns the ultimate fate and longevity of these cells (Malynn et al. 1990). The adult peripheral B cell pool is comprised of a subset (5%–10%) of immature recent bone marrow émigrés, that are characterized by very high expression of heat stable antigen (HSA) and low to negative expression of sIgD (Allman et al. 1992), as well as relatively short-lived cells (2–4 days half-life) and a subset of relatively long-lived cells (4- to 8-week half-life) which ultimately becomes the majority population (Sprent and Tough 1994). Although it appears that only a small minority of newly generated cells enters the long-lived pool, the mechanisms for selection of this population is unknown. Nonetheless, since this selection in some cases appears to be H chain specific (Malynn et al. 1990), and the presence of soluble antigen can dictate a short-lived status for cells whose sIg receptors bind that antigen (e.g., anergic cells) (Cyster et al. 1994), the selection for longevity must also be considered as a mechanism for pruning the ultimately expressed repertoire.

As mentioned above, a fundamental characteristic of adult bone marrow B cell generation is that clonal expansion, encompassing up to six divisions, occurs at the pre-B cell stage and prior to sIg expression (Decker et al. 1991a; Osmond 1991). Thus, once a cell expresses an H chain that can facilitate clonal maturation (presumably by interaction with SL), that cell proliferates, generating a clone of up to 64 cells with identical H chains. Division then ceases and the members of the clone independently express L chains. Thus, a B cell "clone" actually represents a set of cells with shared H chains but each expressing a different L chain and thus antibody clonotype (Caton 1990; Decker et al. 1991a). This process maximizes the utilization of "functional" H chains but also ensures the greatest extent of repertoire diversity. Among B cell subsets, this strategy is uniquely applicable to conventional primary AFC precursors and suggests that maximum diversity per se may be the intended outcome for this subset.

2.2 Repertoire Expression in Neonatal B Cells

Whereas maximum diversification characterizes repertoire generation in the adult bone marrow, the opposite outcome, i.e., the reproducible expression of a limited repertoire appears to be a major aspect of early neonatal B cell generation (Klinman and Press 1975; Sigal et al. 1976; Cancro et al. 1979; Denis and Klinman 1983). Thus, while both repertoires emanate from the recombination of the same sets of V region gene segments, in fetal and neonatal development repertoire diversity is minimized by: (a) the overutilization of certain H and L chain V gene segments (Alt et al. 1984; Perlmutter et al. 1985; Teale and Medina 1992); (b) the absence of terminal deoxynucleotidyl transferase (TdT) and thus a lack of N additions (Feeney 1990; Gu et al. 1990; Hardy et al. 1991); and (c) the overexpression of certain D_H–J_H and V_H–D_H junctions by the favoring of recombination of gene segments with overlapping terminal nucleotide sequences (Feeney 1990; Gu et al. 1990). Because of these factors, the early repertoire is extremely limited and increases in diversity very gradually (Klinman and Press 1975; Sigal et al. 1976; Cancro et al. 1979; Denis and Klinman 1983). Thus, even though there are few B cells in a neonate, the number of clonotypes is far fewer than the total number of B cells (multiple cells per clonotype) and much of the repertoire is reproducibly expressed in a time-ordered fashion by genetically identical individuals (Klinman and Press 1975; Sigal et al. 1976; Cancro et al. 1979; Denis and Klinman 1983; Riley et al. 1986). Added to this it appears likely that clonal expansion occurs not only before but also after sIg expression dictating the presence of multiple cells expressing the same clonotype (Klinman and Press 1975; Sigal et al. 1976; Cancro et al. 1979; Denis and Klinman 1983; Teale and Medina 1992).

Because the repertoire of neonates is so restricted, it may be considered that much of the neonatal repertoire is genetically predetermined. This contrasts with the generation of the adult repertoire which is dominated by stochastic events that create a vast but unpredictable repertoire. Because of the reproducible nature of the neonatal repertoire it has been suggested that many of the expressed specificities may have been evolutionarily selected as crucial for early defense against endemic pathogens or as mediators in regulating further repertoire expression (Chen and Kearney 1993). Additionally, some neonatally generated clonotypes appear to persist in the adult, primarily as B-1 cells (see below), and can serve an important role in antibacterial protection throughout adulthood.

2.3 Repertoire Expression in B-1 Cells

B-1 cells constitute a naive B cell subset that, in adult mice of most strains, is prevalent among peritoneal cells but rare in most other B cell populations and overall represent 2%–5% of all B cells. The numerous unique characteristics of this cell subset have been reviewed extensively elsewhere (HARDY 1993; KANTOR and HERZENBERG 1993; KANTOR et al. 1997). This population predominates in fetal and early neonatal B cell populations and many of these neonatal B-1 cells appear to persist into adulthood. However, for several weeks after birth B-1 cells are generated from the bone marrow. Although these bone marrow generated B-1 (B-1b) cells share many of the characteristics of the earlier generated B-1a cells, including high surface IgM and low sIgD, in general they lack expression of CD-5 which characterizes the earlier generated cells. Most significant, many B-1 cells, especially those that are generated in fetal and early neonatal life, lack N additions. However, with time, more and more B-1 cells are N^+ (KANTOR et al. 1997).

Because of the major contribution of early generated cells, B-1 cells share many of the repertoire characteristics of neonatal B cells, in particular expanded clones and a relatively restricted repertoire. With time, however, this population diversifies but still has marked differences in repertoire expression from conventional primary AFC precursors (HARDY 1993; KANTOR et al. 1997). Most of these differences appear to be the result of the highly selective process by which cells enter the B-1 pool (ARNOLD et al. 1994). While it remains controversial whether cells are destined to become B-1 cells early in pro-pre-B cell development or only after sIg expression, in all cases entrance into the mature B-1 cell pool appears to require interactions involving the cells' sIg receptors. In addition, the signal that enables entrance into the B-1 pool also appears to engender clonal expansion (CLARKE and MCCRAY 1993). This, added to the relatively small total numbers of B-1 cells and the enhanced ability of B-1 cells to self renew in the periphery, results in a B-1 cell repertoire that is relatively restricted and highly skewed towards certain clonotypes. Indeed, the predominant clonotypes characteristic of anti-phosphorylcholine (PC) and anti-phosphatidylcholine responses are largely found in the B-1 population (KANTOR and HERZENBERG 1993). In this regard, the generation and maintenance of the B-1 population differs dramatically from either conventional or neonatal B cells in that clonal maintenance and expansion appears highly sIg-receptor dependent. This yields a far more environmentally selected population than conventional B cells, even though the longevity of conventional AFC precursors may also reflect environmental selection.

3 Repertoire Expression in Memory B Cells

Although the fundamental molecular mechanisms responsible for repertoire establishment are common for memory and naive B cells, the underlying rationale of memory B cell repertoire generation is vastly different from that of: (a) the primary conventional AFC repertoire wherein diversity per se is maximized, (b) the neonatal B cell repertoire wherein expression of expanded clones of evolutionarily conserved clonotypes is ensured, or (c) the B-1 cell repertoire wherein environmentally selected and expanded clones predominate. Like neonatal and B-1 cells, clonal expansion is central to memory B cell development and, in this sense, the presence of memory B cells may reduce overall repertoire diversity. However, the modus operandi of memory repertoire generation is first to access those elements of the naive repertoire that recognize a foreign antigen and then to simultaneously generate expanded clones of these cells and refine their expressed repertoire by successive rounds of somatic mutation and selection. Thus, in contrast to the primary repertoires wherein antigens can only select among pre-existing clonotypes, the memory response can create a new and far more efficient repertoire in response to a recognized antigen.

Memory B cell generation requires T_H and various co-receptor interactions in the initial stimulation event, as well as the germinal center (GC) environment and follicular dendritic cells (FDC) (JACOBSON et al. 1974; LIU et al. 1986; TEW et al. 1990; BEREK et al. 1991; GRAY et al. 1991; JACOB et al. 1991). However, while there is a general consensus concerning the impact of somatic mutation and selection (CLARKE et al. 1985; BEREK and MILSTEIN 1987; RAJEWSKY et al. 1987), many fundamental issues concerning memory B cell generation are poorly understood and others are highly controversial. These issues include the identity of the cells that can initiate memory responses, the nature and diversity of the repertoire of pre-existing specificities available for selection into the memory B cell pool, the parameters of memory B cell stimulation, the mechanism responsible for somatic hypermutation, the basis and mechanism of both positive and negative selection of newly generating memory B cells, and the mechanisms responsible for the propagation and maintenance of B cell memory. Thus, although it is clear that the strategy used for generating the memory B cell repertoire contrasts greatly from that responsible for the generation of each of the naive B cell subsets, as will be reviewed below, our current understanding of the basis for memory B cell generation remains relatively primitive.

3.1 The Preimmune Repertoire Available for Incorporation Into Memory B Cells

A fundamental question that has yet to be resolved is whether the spectrum of preimmune clonotypes that can be incorporated into memory responses overlaps completely with the repertoire expressed in the primary B cell subsets. Somatically mutated memory B cell V regions can have non-mutated counterparts in the conventional naive AFC precursor pool, as well as the neonatal and B-1 B cell pools (PERLMUTTER et al. 1984; CLARKE et al. 1985; BEREK and MILSTEIN 1987; MALIPIERO et al. 1987; MANSER et al. 1987; RAJEWSKY et al. 1987; BLIER and BOTHWELL 1988). Indeed, there are several examples of sIg transgenic mice wherein the transgene-encoded antibody is present in both primary AFC and memory B cells

(Storb 1987; Carsetti et al. 1993; Cyster et al. 1994). Because of this, and the fact that a single T_H dependent antigenic stimulation can give rise to both primary AFC responses and memory B cells that share clonotypic origins, many investigators have concluded that memory B cells arise from the pool of primary AFC precursors, perhaps as the product of unequal division. This was precisely the conclusion of Jacob and Kelsoe (1992) who demonstrated that, in some instances, 10 days after immunization splenic AFC clusters expressed antibodies with unmutated V regions that arose from the same progenitor that originated the B cells with mutated V regions in proximal GC. Thus, they suggest that, in these instances, the precursor that gave rise to a primary AFC clone also initiated the generation of memory B cells.

In contrast to this conclusion our laboratory, and others, have found that peripheral B cell populations of naive mice, fractionated on the basis of their low surface heat stable antigen (HSA^{lo}) expression, are greatly enriched for cells that can give rise to memory B cells but yield little, if any, primary AFC (Linton et al. 1989, 1992; Yin and Vitetta 1992; Wu and Ward 1993; Decker et al. 1995a). Conversely, cells with intermediate to high HSA levels ($HSA^{int\text{-}hi}$) yield AFC responses but not memory B cells. Importantly, the progeny of HSA^{lo} but not $HSA^{int\text{-}hi}$ cells accumulate somatic mutations after in vivo or in vitro stimulation (Linton et al. 1989; Decker et al. 1995) and also give rise to GC upon transfer with T_H to irradiated severe combined immunodeficiency (SCID) recipients (Linton et al. 1992). The findings that have led to these conclusions are reviewed extensively elsewhere (Klinman and Linton 1990; Klinman 1996b). Relevant to the issue of the origins and extent of repertoire diversity in the cells that give rise to memory B cells, the conclusion that a separate subset comprising only perhaps10% of all naive B cells gives rise to memory responses raises several intriguing questions. Firstly, if only 10% of cells can give rise to memory responses, the repertoire of these cells should be more restricted than that of primary responses. To the contrary, memory responses have generally been found to be highly diverse, and in many situations, appear to be derived from clonotypes that are rare or absent among primary AFC responses (Kaplan et al. 1985; Press and Giorgetti 1986, 1993; Jemmerson 1987; Duran and Metcalf 1987; Caton et al. 1996). Indeed, although in many instances the same clonotypes can be found in primary AFC and memory responses, the finding that certain antigenic determinants are recognized only by memory B cells and numerous clonotypes are unique to memory responses is highly suggestive of a distinct B cell subset originating memory responses. Consistent with the conclusion that memory responses may incorporate clonotypes that are not present in the naive conventional AFC precursor pool is the finding that the TEPC-15 clonotype, which dominates the B-1 cell response to PC, but which is rarely found in conventional AFC precursors (Kantor and Herzenberg 1993), can be found among somatically mutated anti-PC memory B cells (Perlmutter et al. 1984; Malipiero et al. 1987; Wu and Ward 1993). Since B-1 cells do not generally give rise to GC or somatically mutated memory B cells, these findings suggest that the TEPC-15 clonotype is present in the populations of progenitors of memory B cells and B-1 cells but not conventional AFC precursors.

It should also be noted that, in addition to the contribution of B-1 cells to primary AFC clonotypes that can overlap with memory clonotypes, the progeny of memory precursors per se could also contribute to apparent overlaps of primary AFC and memory clonotypes. Thus, rather than being the product of primary conventional AFC precursors, AFC whose V regions are unmutated can be generated via restimulation by persistent antigen of newly generated memory B cells (LINTON et al. 1989; DECKER et al. 1995a). Numerous investigators have documented the lack of V region somatic mutations during the first several days of memory B cell generation (MALIPIERO et al. 1987; BEREK et al. 1991; JACOB et al. 1991). Importantly, these newly generated non-mutated memory B cells can be restimulated by antigen and T_H to give rise to AFC clones producing non-mutated antibodies (LINTON et al. 1989; DECKER et al. 1995a). Thus, towards the end of the first week of an immune response, the unmutated progeny of memory B cells may contribute significantly to the AFC pool. Indeed, the restimulation of GC generated cells could readily account for the non-mutated AFC that share clonotype with proximal GC cells observed by JACOB and KELSOE (1992) (see above). Because of this, there remains no compelling evidence that primary AFC precursors (either conventional or B-1) can simultaneously give rise to clones of AFC and memory B cells.

How is it possible for memory responses to include a broader spectrum of clonotypes than primary AFC responses, especially if far fewer cells can give rise to memory B cells? One explanation for this could be the possibility that memory responses could be induced from cells whose sIg receptors were not only high affinity for the antigen (like primary AFC precursors), but also those with much lower affinity. It has been suggested that the inclusion of cells with lower affinity receptors could readily account for the presence of clonotypes unique to memory responses, particularly since somatic mutation and selection could rapidly convert these V regions to high affinity (PRESS and GIORGETTI 1986, 1993). Such a hypothesis would be consistent with the finding that the frequency of responsive HSA^{lo} memory progenitors is substantially higher than the frequency of $HSA^{int\text{-}hi}$ primary AFC precursors on a per-cell basis (LINTON et al. 1986; KLINMAN 1996b). Additionally, naive progenitors of memory B cells, like memory B cells per se, can be more easily stimulated by cross-reactive (low affinity) antigens than can primary AFC precursors (KLINMAN et al. 1973, 1974; LINTON and KLINMAN, unpublished observation).

3.2 The Propagation of Memory Responses

An extremely important issue which has received relatively little attention concerns the cells that are responsible for propagating B cell memory. Because memory B cells and somatic mutations appear to accumulate as a consequence of the GC reaction, the inability to readily observe GC formation from memory B cells has been interpreted as indicating that higher order memory responses (tertiary, quaternary, etc, responses) must emanate from memory B cells generated during the initial priming event or from memory B cells generated anew upon each subsequent immunization (LIU et al. 1986; GRAY et al. 1991). This hypothesis received support from findings of SIEKEVITZ et al. (1987) that indicated that, upon transfer and stimulation in adoptive hosts, somatically mutated memory B cells generated AFC clones without the significant accumulation of more mutations.

The opposite conclusion has been reached by investigators assessing the number and distribution of somatic mutations within V regions of B cells after sequential immunizations (CLARKE et al. 1985; BEREK and MILSTEIN 1987; BEREK and ZIEGNER 1993). The data from such studies have been interpreted as indicating, not only that there is an increase in the average number of mutations with subsequent immunizations, but also that there is an accumulation of new mutations within already mutated V regions. Such findings clearly suggest that, upon stimulation, mutated memory B cells must be able to undergo further rounds of memory B cell generation accompanied by the further accumulation of somatic mutations. If this is the case, then the inability to readily identify GC formation from memory B cells may either reflect the ability of such cells to generate new memory B cells and somatic mutations without any, or only evanescent GC formation, or that the experimental systems used to assess GC formation from memory B cells were inadequate.

Recent findings from in vitro analyses of the potential of various murine B cell subsets to generate somatically mutated memory B cells have suggested an alternative hypothesis for the generation of higher order memory B cells (DECKER et al. 1995; KLINMAN 1996b). In these fragment culture studies, as anticipated, naive HSA^{lo} B cells gave rise to memory B cell clones with GC markers after in vitro stimulation, and the AFC generated after restimulation of these cells had accumulated a significant number of somatic mutations. Also, not surprisingly, the AFC that were generated after a single stimulation of splenic B cells had no V region somatic mutations.

Thus, the ability of cells to somatically mutate could be assessed at the clonal level in vitro and the capacity to somatically mutate was found only in the progeny of cells that generated memory B cells. Surprisingly, a similar analysis of the population of HSA^{lo}, isotype-switched, memory B cells obtained from previously immunized mice also revealed two distinct response phenotypes. As anticipated, most memory B cells generated a vigorous AFC response after a single 2-day course of in vitro antigenic stimulation. Sequence analysis of V regions from the AFC of these clones confirmed the memory phenotype of the cells that originated the clone in that they contained numerous somatic mutations. However, consistent with the aforementioned studies of SIEKEVITZ et al. (1987), all sequences obtained from a single clone were identical, indicating that there was no further accumulation of mutations during AFC generation after antigenic stimulation of memory B cells.

A minority of antigen responsive memory B cells displayed a vastly different response phenotype. These cells generated no antibody responses after a single course of antigenic stimulation. However, after restimulation at days 8–10 of culture, vigorous AFC responses were obtained. Thus, the response phenotype of this subset of memory B cells was very much like that previously observed for

HSA^{lo} naive B cells. These findings have been interpreted as indicating that, upon stimulation, this subset of memory precursors generated clones of higher order (tertiary, etc.) memory B cell clones which require restimulation to generate AFC. Importantly, the memory cell status of these "immune memory progenitors" was evidenced not only by their switched surface Ig isotype but also by the fact that all sequences obtained from their AFC progeny shared numerous somatic mutations, indicating that the cell that originated the clone was already mutated. However, many of the sequences also showed unique mutations as well, indicating that this population retained the capacity to further accumulate somatic mutations. We believe the existence of a subset of memory B cells that can propagate further generations of memory B cells and continue to accumulate somatic mutations resolves much of the above controversy and provides a cellular rationale for the aforementioned incremental accumulation of somatic mutations upon subsequent immunizations.

3.3 Positive and Negative Selection in Memory B Cell Generation

It is now generally accepted that affinity maturation in memory B cells is the consequence of powerful affinity-driven selective forces acting on cells whose sIg receptors are rapidly evolving through the accumulation of somatic mutations (Clarke et al. 1985; Berek and Milstein 1987; Manser et al. 1987; Rajewsky et al. 1987; Blier and Bothwell 1988). The extent of this affinity driven molecular evolution varies from response to response in that in some responses the initial affinity of unmutated receptors is not improved by mutations (Perlmutter et al. 1984; Malipiero et al. 1987; Zinkernagel 1996), while in other responses mutations improve affinity dramatically (Clarke et al. 1985; Berek and Milstein 1987; Rajewsky et al. 1987). In the latter cases, affinity maturation and the accumulation of somatic mutations is so marked that selection must be both intensive and applicable at most, if not all cell divisions. Given the remarkable impact of positive selection on the repertoire of memory B cells, it is surprising that so little is understood about the selective mechanisms per se. Thus, while it is thought that antigen presented by the iccosomes of FDC (Tew et al. 1990) and the ability to capture T cell help play a central role in affinity maturation, it remains a matter of speculation whether these or sIg receptor-mediated interactions propagate mutations or shut them off (Shlomchik et al. 1989; Berek and Zeigner 1993). Furthermore, there is presently no coherent concept concerning the balance between the increase in responsive cells due to proliferation vs the loss in responsive cells due to deleterious mutations, tolerance, and the terminal differentiation of cells to AFC or AFC precursors.

In contrast to the dearth of information concerning the positive selection of newly generating memory cells that accounts for the accumulation of mutations and affinity maturation, there is a growing body of information concerning the negative selection of newly arising memory B cells. Because of the generation of new specificities due to V region somatic mutations, the possibility that newly generating memory B cells may become reactive to self-antigens has long been recognized. The existence of a "second window" of tolerance susceptibility for newly generating memory B cells was demonstrated several years ago, through the analysis of memory B cells generated in fragment cultures (LINTON et al. 1991). In this case, tolerance susceptibility of newly generating B cells had many of the hallmarks of tolerance susceptibility of newly generating adult bone marrow and neonatal B cells in that multivalent interactions (receptor crosslinking) was required and concomitant T cell help could yield stimulation rather than tolerance (METCALF and KLINMAN 1976, 1977; TEALE and KLINMAN 1980, 1984; RILEY and KLINMAN 1986; LINTON et al. 1991; KLINMAN 1996). Interestingly, unlike neonatal and immature bone marrow cells, in the case of memory B cells, cross-reactive antigens were highly potent tolerogens. Thus, cells of the memory compartment appeared to have a lower affinity threshold for sIg receptor-mediated tolerance induction.

Recently, several laboratories have used in vivo analyses to confirm the tolerance susceptibility of newly generating GC B cells (HAN et al. 1995; PULENDRAN et al. 1995; SHOKAT and GOODNOW 1995). In these studies, the addition of high concentrations of tolerogen to mice undergoing a GC reaction markedly increased the number of apoptotic GC B cells and decreased GC size.

4 Concluding Remarks

Primary B cell subsets must provide pre-existing repertoires that are both sufficiently diverse to provide protection against an extremely broad set of pathogens, including those which may be novel to the species, and at the same time ensure reproducible recurrence of cells whose V region clonotypes may be crucial for defense against evolutionarily important pathogens. In contrast, the memory B cell repertoire relies only on the availability of a few reactive cells, possibly even those with low affinity for pathogens, and uses the process of T_H dependent antigenic stimulation to simultaneously increase the number of responsive cells, increase the diversity of reactive cells, and select among the new specificities those that have the highest affinity for the pathogen, while being non-reactive to self antigens.

We suggest that the distinct functional roles that the primary vs memory B cell repertoires must subserve have profound implications not only for the means by which these repertoires are generated, but also for the parameters of both their stimulation and tolerance induction. Because of the need for the preservation of both enormous diversity and recurrent clonotypes, tolerance induction of newly generating primary B cells is highly affinity-dependent and selective. Thus, while cells bearing high-affinity anti-self specificities are eliminated, the selection process is apparently permissive for cells with lower affinity anti-self, thus preserving the most diverse possible repertoire. Conversely, during memory B cell generation an enormous capacity for increased diversity of clonotypes specific for the immunogen can be superimposed on the pre-existing repertoire. Thus, even very low affinity anti-self specificities can be eliminated without impeding the generation, selection, and propagation of a highly efficient population of memory B cells.

References

Allman DM, Ferquson SE, Cancro MP (1992) Peripheral B cell maturation. I. Immature peripheral B cells in adults are heat-stable antigen^{hi} and exhibit unique signaling characteristics. J Immunol 149:2533–2540

Alt FW, Yancopoulos GD, Blackwell TK, Wood C, Thomas E, Boss M, Coffman R, Rosenberg N, Tonegawa S, Baltimore D (1984) Ordered rearrangement of immunoglobulin heavy-chain variable region segments. EMBO J 3:1209–1219

Arnold LW, Pennell CA, McCray SK, Clarke SH (1994) Development of B-1 cells: segregation of phosphatidylcholine-specific B cells to the B-1 population occurs after immunoglobulin gene expression. J Exp Med 179:1585–1595

Berek C, Milstein C (1987) Mutation drift and repertoire shift in the maturation of the immune response. Immunol Rev 96:23–41

Berek C, Ziegner M (1993) The maturation of the immune response. Immunol Today 14:400–404

Berek C, Beige A, Apel M (1991) Maturation of the immune response in germinal centers. Cell 67:1121–1129

Blier PR, Bothwell ALM (1988) The immune response to the hapten NP in C57BL/6 mice: insights into the structure of the B cell repertoire. Immunol Rev 105:27–43

Blomberg B, Geckeler WR, Weigert M (1972) Genetics of the antibody response to dextran in mice. Science 177:178–180

Cancro MP, Gerhard W, Klinman NR (1978) The diversity of the primary influenza specific B cell repertoire in BALB/c mice. J Exp Med 147:776–782

Cancro MP, Wylie DE, Gerhard W, Klinman NR (1979) Patterned acquisition of the antibody repertoire: diversity of the hemaglutinin-specific B cell repertoire in neonatal BALB/c mice. Proc Natl Acad Sci USA 76:6577–6581

Carsetti R, Kohler G, Lamers MC (1993) A role for immunoglobulin D: interference with tolerance induction. Eur J Immunol 23:168–178

Caton AJ (1990) A single pre-B cell can give rise to antigen-specific B cells that utilize distinct immunoglobulin gene rearrangements. J Exp Med 172:815–825

Caton AJ, Swartzentruber JR, Kuhl AL, Carding SR, Stark SE (1996) Activation and negative selection of functionally distinct subsets of antibody-secreting cells by influenza hemagglutinin as a viral and a neo-self antigen. J Exp Med 183:13–26

Chen X, Kearney JF (1993) V_H81X encoded IgM transgenic mice. J Immunol 150:253A (Abstract 1444)

Clarke SH, McCray SK (1993) V_HCDR3-dependent positive selection of murine V_H12-expressing B cells in neonates. Eur J Immunol 23:3327–3334

Clarke SH, Huppi K, Ruezinsky D, Staudt L, Gerhard W, Weigert M (1985) Inter and intraclonal diversity in the antibody response to influenza hemagglutinin. J Exp Med 161:687–704

Cooper HM, Klinman NR, Paterson Y (1988) The auto-antigenic response to rabbit cytochrome c. Eur J Immunol 19:315–322

Crews S, Griffin J, Huang H, Calame K, Hood L (1981) A single VH gene segent encodes the immune response to phosphorylcholine: somatic mutation is correlated with the class of the antibody. Cell 25:59–66

Cyster JG, Hartley SB, Goodnow CC (1994) Competition for follicular niches excludes self-reactive cells from recirculating B cell repertoire. Nature (Lond) 371:389–395

Decker DJ, Boyle NE, Koziol J, Klinman NR (1991a) The expression of the immunoglobulin heavy chain repertoire in developing bone marrow B lineage cells. J Immunol 146:350–361

Decker DJ, Boyle NE, Klinman NR (1991b) Predominance of non-productive rearrangements of V_H81X gene segments evidences a dependence of B cell clonal maturation on the structure of nascent H chains. J Immunol 147:1406–1411

Decker DJ, Linton PJ, Jacobs SN, Biery M, Gingeras TR, Klinman NR (1995a) Defining subsets of naive and memory B cells based on their ability to somatically mutate vitro. Immunity 2:195–203
Decker DJ, Kline GH, Hayden TA, Zaharevitz SN, Klinman NR (1995b) Heavy chain V gene-specific elimination of B cells during the pre B cell to B cell transition. J Immunol 154:4925–4935
Denis KA, Klinman NR (1983) Genetic and temporal control of neonatal antibody expression. J Exp Med 157:1170–1183
Desiderio SV, Yancopoulos GD, Paskind E, Thomas E, Boss M, Landau N, Alt FW, Baltimore D (1984) Insertion of N regions into heavy-chain genes is correlated with expression of terminal deoxytransferase in B cells. Nature 311:752–755
Duran LW, Metcalf ES (1987) Clonal analysis of primary B cells responsive to the pathogenic bacterium Salmonella typhimurium. J Exp Med 165:340–358
Early P, Huang H, Davis M, Calame K, Hood L (1980) An immunoglobulin heavy chain variable region gene is generated from three segments of DNA, V_H, D, and J_H. Cell 19:981–992
Feeney AJ (1990) Lack of N-regions in fetal and neonatal mouse immunoglobulin V-D-J junctional sequences. J Exp Med 172:1377–1390
Feeney AJ, Riblet R (1993) DST4: a new, and probably the last, functional D_H gene in the BALB/c mouse. Immunogenetics 37:217–221
Fitts MG, Mage RG (1995) Secondary rearrangements and post-rearrangement selection contribute to restricted immunoglobulin D-J_H expression in young rabbit bone marrow. Eur J Immunol 25:700–707
Froscher BG, Klinman NR (1985) Strain-specific silencing of a predominant antidextran clonotype family. J Exp Med 162:1620–1633
Goodnow CC (1992) Transgenic mice and analysis of B cell tolerance. Annu Rev Immunol 19:489–518
Gray D, Kosco M, Stockinger B (1991) Novel pathways of antigen presentation for maintenance of memory. Int Immunol 3:141–148
Gu H, Forster I, Rajewsky K (1990) Sequence homologies, N sequence insertion and J_H gene utilization in V_H–D_H–J_H joining: implications for the joining mechanism and the ontogenic timing of Ly1 B cell and B-CLL progenitor generation. EMBO J 9:2133–2140
Han S, Zheng B, Dal Porto J, Kelsoe G (1995) In situ studies of the primary immune response to (4-hydroxy-3-nitrophenyl)acetyl IV. Affinity-dependent, antigen-driven B cell apoptosis in germinal centers as a mechanism for maintaining self-tolerance. J Exp Med 182:1635–1644
Hardy RR (1993) Variable gene usage, physiology and development of Ly-1$^+$ (CD5$^+$) B cells. Curr Opin Immunol 4:181–185
Hardy RR, Carmack CE, Shinton SA, Kemp JD, Hayakawa K (1991) Resolution and characterization of pre-pro B cell stages in normal mouse bone marrow. J Exp Med 173:1213–1225
Harris DE, Cairns L, Rosen FS, Borel Y (1982) A natural model of immunologic tolerance. Tolerance to murine C5 is mediated by T cells and antigen is required to maintain unresponsiveness. J Exp Med 156:567–584
Hartley SB, Cooke MP, Fulcher DA, Harris AW, Cory S, Basten A, Goodnow CC (1993) Elimination of self-reactive B lymphocytes proceeds in two stages: arrested development and cell death. Cell 72:325–335
Huetz F, Carlsson L, Tornberg UC, Holmberg D (1993) V region directed selection in differentiating B lymphocytes. EMBO J 12:1819–1826
Ichihara Y, Hayashida H, Miyazawa S, Kurosawa Y (1989) Only D_{FL16}, D_{SP2}, and D_{Q52} gene familes exist in mouse immunoglobulin heavy chain diversity gene loci, of which D_{FL16} originate from the same primordial D_H gene. Eur J Immunol 19:1849–1854
Jacob J, Kelsoe G (1992) In situ studies of the primary immune response to (4-hydroxy-3-nitrophenyl)acetyl. II. A common clonal origin for periarteriolar lymphoid sheath asssociated foci and germinal centers. J Exp Med 176:679–687
Jacob J, Kelsoe G, Rajewsky K, Weiss U (1991) Intraclonal generation of antibody mutants in germinal centers. Nature 354:389–392
Jacobson EB, Caporale LH and Thorbecke GJ (1974) Effect of thymus cell injections on germinal center formation in lymphoid tissues of nude (thymusless) mice. Cell Immunol 13:416–430
Jemmerson RW (1987) Multiple overlapping epitopes in the three antigenic regions of horse cytochrome c. J Immunol 138:213–219
Kantor AB, Herzenberg LA (1993) Origin of murine B cell lineages. Annu Rev Immunol 11:501–538
Kantor AB, Merrill CE, Herzenberg LA, Hillson JL (1997) An unbiased analysis of V_H-D-J_H sequences from B-1a, B-1b, and conventional B cells. J Immunol 158:1175–1186
Kaplan MA, Ching LK, Berte C, Sercarz EE (1985) Predominant idiotypy and specificity shift during the antibody response to lysozyme (abstract). Fed Proc 44:1692

Kenya U, Beck-Engeser GB, Jongstra J, Applequist SE, Jäck HM (1995) Surrogate light chain dependent selection of immunoglobulin heavy chain variable regions. J Immunol 155:5536–5542
Kitamura D, Kudo A, Schaal S, Muller W, Melchers F, Rajewsky K (1992) A critical role of l-5 protein in B cell development. Cell 69:823–831
Klinman NR (1996a) The "clonal selection hypothesis" and current concepts of B cell tolerance. Immunity 5:189–195
Klinman NR (1996b) In Vitro analysis of the generation and propagation of memory B cells. Immunol Rev 150:91–111
Klinman NR, Linton PJ (1990) The generation of B cell memory: a working hypothesis. In: Sprent J, Gray D (eds) Current topics in microbiology and immunology, vol 159. Springer, Berlin Heidelberg New York, pp 19–35
Klinman NR, Press JL (1975) The characterization of the B cell repertoire specific for the 2,4-dinitrophenyl and 2,4,6-trinitrophenyl determinants in neonatal BALB/c mice. J Exp Med 141:1133–1146
Klinman NR, Stone MR (1983) Role of variable region gene expression and environmental selection in determining the anti-phosphorylcholine B cell repertoire. J Exp Med 158:1948–1961
Klinman NR, Press JL, Segal G (1973) Overlap stimulation of primary and secondary B cells by cross-reacting determinants. J Exp Med 138:1276–1281
Klinman NR, Press JL, Pickard AR, Woodland RT, Dewey AF (1974) The biography of the B cell. In: Sercarz E, Williamson A, Fox CF (eds) The immune system. Academic, New York, pp 357–365
Klinman NR, Kline GH, Hartwell L, Beck-Engesser G, Keyna U, Jäck HM (1997) Heavy chain assembly with surrogate light chain is required for allelic exclusion and pre B cell maturation (abstract). Keystone Symposia on Molecular and Cellular Biology: B lymphocytes in health and disease, p 6
Kurosawa Y, Tonegawa S (1982) Organization, structure and assembly of immunoglobulin heavy chain diversity DNA segments. J Exp Med 155: 201–218
Lieberman R, Potter M, Mushinski W, Humphrey W, Rudikoff S (1974) Genetics of a new IgV_H (T15 idiotype) marker in the mouse regulating natural antibody to phosphorylcholine. J Exp Med 139:983–1001
Linton PJ, Decker DJ, Klinman NR (1989) Primary antibody forming cells and secondary B cells are generated from separate precursor cell subpopulations. Cell 59:1049–1059
Linton PJ, Rudie A, Klinman NR (1991) Tolerance susceptibility of newly generating memory B cells. J Immunol 146:4099–4104
Linton PJ, Lo D, Lai L, Thorbecke GJ, Klinman NR (1992) Among naive precursor cell subpopulations only progenitors of memory B cells originate germinal centers. Eur J Immunol 22:1293–1297
Liu YJ, Mason DY, Johnson GD, Abbot S, Gregory CD, Hardie DL, Gordon J, MacLennan ICM (1986) Germinal center cells express bcl-2 protein after activation by signals which prevent their entry into apoptosis. Eur J Immunol 21:1905–1910
Makela O, Karajalainen K (1977) Inherited immunoglobulin idiotypes of the mouse. Immunol Rev 345:119–138
Malipiero UV, Levy NS, Gearhart PJ (1987) Somatic mutation in anti-phosphorylcholine antibodies. Immunol Rev 96:59–74
Malynn BA, Yancopoulos GD, Barth JE, Bona A, Alt FW (1990) Biased expression of J_H-proximal V_H genes occurs in the newly generated repertoire of neonatal and adult mice. J Exp Med 171:843–859
Manser T, Wysocki LJ, Margolies MN, Gefter ML (1987) Evolution of antibody variable region structure during the immune response. Immunol Rev 96:141–162
Max EE (1984) Immunoglobulins: molecular genetics. In: Paul WE (ed) Fundamental immunology. Raven, New York, pp 167–204
Metcalf ES, Klinman NR (1976) In vitro tolerance induction of neonatal murine B cells. J Exp Med 143:1327–1386
Metcalf ES, Klinman NR (1977) In vitro tolerance induction of bone marrow cells: a marker for B cell maturation. J Immunol 118:2111–2116
Metcalf ES, Schrater AF, Klinman NR (1978) Murine models of tolerance induction in developing and mature B cells. Immunol Rev 43:143–183
Milstein C, Even J, Jarvis JM, Gonzales-Fernandez A, Gherardi E (1992) Non-random features of the repertoire expressed by the members of one V kappa gene family and of the V-J recombination. Eur J Immunol 22:1958–1962
Nemazee DA, Burki K (1989) Clonal deletion of B lymphocytes in a transgenic mouse bearing anti-MHC class I antibody genes. Nature 337:562–566
Nossal GJV (1983) Cellular mechanisms of immunologic tolerance. Annu Rev Immunol 1:33–62

Osmond DG (1991) Proliferation kinetics and the lifespan of B cells in central and peripheral lymphoid organs. Curr Opin Immunol 3:179–185
Park YH, Osmond DG (1989) Dynamics of early B lymphocyte precursor cells in mouse bone marrow: proliferation of cells containing terminal deoxynucleotidyl transferase. Eur J Immunol 19:2139–2144
Pawlak LL, Nisonoff A (1973) Distribution of a crossreactive idiotypic specificity in inbred strains of mice. J Exp Med 1139:869–983
Perlmutter RM, Crews ST, Douglas R, Sorensen G, Johnson N, Nivera N, Gearhart PJ, Hood L (1984) The generation of diversity in phosphorylcholine-binding antibodies. Adv Immunol 35:1–37
Perlmutter RM, Kearney JF, Chang SP, Hood LP (1985) Developmentally controlled expression of immunoglobulin V_H genes. Science 227:1597–1601
Press JL, Giorgetti CA (1986) Clonal analysis of the primary and secondary B cell responses of neonatal, adult, and xid mice. J Immunol 139:608–618
Press JL, Giorgetti CA (1993) Molecular and kinetic analysis of an epitope-specific shift in the B cell memory response to a multideterminant antigen. J Immunol 151:1998–2013
Pulendran B, Kannourakis G, Nouri S, Smith KGC, Nossal GJV (1995) Soluble antigen can cause enhanced apoptosis of germinal-center B cells. Nature 375:331–334
Radic MZ, Erickson J, Litwin S, Weigert M (1993) B lymphocytes may escape tolerance by revising their antigen receptors. J Exp Med 177:1165–1173
Rajewsky K, Forster L, Cumano A (1987) Evolutionary and somatic selection of the antibody repertoire in the mouse. Science 238:1088–1094
Reth M, Jackson S, Alt F (1986) V_H-D-J_H formation and D-J_H replacement during pre B differentiation: non-random usage of gene segments. EMBO J 5:2131–2138
Riley RL, Klinman NR (1986) The affinity threshold for antigenic triggering differs for tolerance susceptible immature precursors vs mature primary B cells. J Immunol 136:3147–3154
Riley SR, Connors SJ, Klinman NR, Ogata RT (1986) Preferential expression of variable region heavy chain gene segments by predominant dinitrophenyl-specific BALB/c neonatal antibody clonotypes. Proc Natl Acad Sci USA 83:2589–2593
Rolink A, Karasuyama H, Grawunder U, Haasner D, Kudo A, Melchers F (1993) B cell development in mice with a defective l-5 gene. Eur J Immunol 23:1284–1288
Russell DM, Dembic Z, Morahan G, Miller JF, Burki K, Nemazee D (1991) Peripheral deletion of self-reactive B cells. Nature 354:308–311
Shlomchik MJ, Litwin S, Weigert M (1989) The influence of somatic mutation on clonal expansion. In: Melchers F (ed) Progress in immunology, VII. Springer, Berlin Heidelberg New York, pp 415–423
Shokat KM, Goodnow CC (1995) Antigen-induced B cell death and elimination during germinal center immune responses. Nature 375:334–343
Siekevitz M, Kocks C, Rajewsky K, Dildrop R (1987) Analysis of somatic mutation and class switching in naive and memory B cells generating adoptive primary and secondary responses. Cell 48:757–770
Sigal NH, Klinman NR (1978) The B cell clonotype repertoire. Adv Immunol 26:225–337
Sigal NH, Gearhart PJ, Press JL, Klinman NR (1976) The late acquisition of a "germ line" antibody specificity. Nature 251:51–52
Sprent J, Tough D (1994) Lymphocyte lifespan and memory. Science 265:1395–1400
Stockinger B, Hausmann B (1988) Induction of an immune response to self-antigen. Eur J Immunol 18:249–253
Storb U (1987) Transgeneic mice with immunoglobulin genes. Annu Rev Immunol 5:151–174
Teale JM, Klinman NR (1980) Tolerance as an active process. Nature 288:385–387
Teale JM, Klinman NR (1984) Membrane and metabolic requirements for tolerance induction of neonatal B cells. J Immunol 133:1811–1817
Teale JM, Medina CA (1992) Comparative expression of adult and fetal V gene repertoires. Int Rev Immunol 8:95–111
Tew JG, Kosco MH, Burton GF, Szakal AK (1990) Follicular dendritic cells as accessory cells. Immunol Rev 117:185–211
Tiegs SL, Russell DM, Nemazee D (1993) Receptor editing in self-reactive bone marrow B cells. J Exp Med 177:1009–1020
Tonegawa S (1983) Somatic generation of antibody diversity. Nature 302:575–581
Wu P, Ward RE (1993) Ig repertoire expression of BALB/c primary and secondary B cell precursors specific for phosphorylcholine. J Immunol 150:3862–3872
Yin XM, Vitetta ES (1992) The lineage relationship between virgin and memory B cells. Int Immunol 4:691–698
Zinkernagel RM (1996) Immunology taught by viruses. Science 271:173–178

Plasticity Under Somatic Mutation in Antigen Receptors

T.B. KEPLER[1] and S. BARTL[2]

1 Introduction

After antigen binding, a subset of newly activated B and T cells is recruited into the primary follicles of lymphoid tissue where, along with follicular dendritic cells, they establish a site of vigorous lymphocyte proliferation. Within these germinal centers the processes of somatic hypermutation and selection result in the affinity maturation of antigen receptors (see, e.g., KELSOE 1996; WAGNER and NEUBERGER 1996). Mutation rates at the rearranged V(D)J locus and the flanking introns are up to 10^6 times higher than background. During the germinal center reaction, the affinities of immunoglobulins (Igs) for the eliciting antigen typically increase by factors of ten or 100 in the B cells that go on to become memory cells. Using antigen as a limited resource in a miniaturized Darwinian struggle for survival among lymphocytes is an essential component of the overall strategy employed by the immune system to effectively combat the vast majority of pathogens. Somatic mutation – though not necessarily accompanied by affinity maturation – has been

[1]Biomathematics Program, Department of Statistics, North Carolina State University, Raleigh, NC 27695-8203, USA
[2]Department of Biological Sciences, University of North Carolina at Wilmington, Wilmington, NC 28403, USA

described in many species, including, remarkably, sharks (HINDS-FREY et al. 1993), the earliest extant animal with an adaptive immune system (DU PASQUIER 1993).

The molecular mechanism of hypermutation remains unknown, but it leaves its signature in a clear sequence specificity. For example, the triplet AGC which, when in frame, encodes serine, is frequently found to be a "hotspot" of hypermutation (BETZ et al. 1993). The statistical analysis of ROGOZIN and KOLCHANOV (1992) suggested that many hotspots are located in one of two motifs: RGYW and TAA, where R = A or G, Y = C or T, and W = A or T. Note that the serine AGY motifs can be consistent with the first of these motifs, but are not invariably so. Although a systematic phylogenetic comparison has not been carried out, it appears that the sequence specificity of the hypermutation mechanism is roughly conserved across a wide variety of organisms, including sharks (HINDS-FREY et al. 1993), toads (WILSON et al. 1992), mice (BETZ et al. 1993), sheep (REYNAUD et al. 1995), humans (VAN DER STOEP et al. 1993), as well as in cell lines (DENÉPOUX et al. 1997) and the T-cell receptor (TCR) variable-region genes of mouse germinal center T cells (ZHENG et al. 1994).

In this chapter, our interests lay not in the nature of the sequence specificity but in the simple fact that such a specificity exists at all. Sequence specificity, together with the redundancy of the genetic code, allows genes that encode identical amino acid sequences to have rather different properties under somatic hypermutation, as has been pointed out by several authors (MOTOYAMA et al. 1991; VARADE et al. 1993; KEPLER, unpublished report; WAGNER et al. 1995). Thus, where a serine residue is to be used in a structurally important framework region (FR), it would be advantageous to encode it using one of the relatively non-mutable serine codons CTN. Conversely, when serine is to be used in the complementarity-determining regions (CDR) the more mutable AGC codon would facilitate hypermutation-driven exploration of the genetic space. Note that we have here a bona fide case of group selection; although it may not be to an individual B cell's advantage to use mutable codons, the host organism, i.e., the target of evolutionary selection, does gain an advantage. Thus, site-specific codon bias could enhance mutability where it would be most beneficial (CDRs) and diminish mutability at sites where it would be harmful (FRs).

Codon bias in general has been recognized for many years (GRANTHAM et al. 1980) and, more recently, evidence for site-specific codon biases in bacteria has been presented (MAYNARD-SMITH and SMITH 1996), but specific mechanisms for the maintenance of site-specific biases in these cases remain unclear.

The use of site-specific codon bias to enhance Ig to focus somatic mutation to CDRs seems straightforward, but its effect may be too subtle. The advantage gained through targeting mutability might not provide an advantage large enough to overcome the substantial stochasticity inherent in evolutionary change. On the other hand, it is difficult to know just what the basis for evolutionary selection of Ig V genes is (see discussions in MÖLLER 1990; ROTHENFLUH et al. 1995; STEWART and COUTINHO 1996). Each gene is but one out of dozens of similar genes in the germline of each organism. Furthermore, several stochastic modifications, including V(D)J rearrangement (reviewed in LEWIS 1994) and random association of heavy- and light-chain genes, as well as somatic mutation, interpose themselves between the germline genes and the expressed Ig repertoire.

In a closely related matter, CHANG and CASALI (1994) found evidence that CDR1 in human Ig V_H genes are inherently susceptible to replacement mutations. That is, CDR1 was found to contain amino acids that are encoded less redundantly than chance would dictate; the R:S ratio expected in the absence of selection (and under a model for somatic hypermutation that ignores sequence specificity) is higher in CDR1 than in FRs. One cannot conclude from this that Ig genes have been selected for this property, since selection at the level of the amino acids themselves rather than the level of the DNA cannot be ruled out.

TANAKA and NEI (1989) analyzed the synonymous vs. non-synonymous substitution rates in murine and human Ig V_H, and reported evidence for diversity-enhancing selection in the CDRs as had been previously found for MHC loci by HUGHES and NEI (1988). These studies found that selection acts to enhance useful diversity in the germline. We, in contrast, seek to determine whether selection acts to enhance diversifiability – *plasticity* – under somatic mutation.

In fact, we find that it does (KEPLER 1997); human V genes do utilize region-specific codon bias to enhance their plasticity under somatic hypermutation. WAGNER et al. (1995) provided evidence for the preferential use of mutable serine codons (AGY) in CDRs and of non-mutable serine codons (CTN) in FRs (although as we will discuss below, there were statistical problems with their analysis). In the first part of this chapter we review our analysis of Ig V gene site-specific codon bias and plasticity. In the second, we apply our methods to the question of hypermutation in TCRs.

Evidence for hypermutation of TCR genes has been found in germinal center T cells (ZHENG 1994), but these findings remain controversial (BACHL and WABL 1995; KELSOE et al. 1995). There are good reasons for circumspection with regard to claims of TCR hypermutation. Unlike Igs that bind antigens alone, TCRs bind antigen (as peptides) in the context of self-encoded MHC molecules. TCR hypermutation would appear to greatly increase the risk of autoimmunity. Nevertheless, if somatic hypermutation plays (or did once play) a functionally significant role in diversifying TCRs, we might detect a site-specific codon bias similar to that found in Ig.

Thus, having developed the statistical methods for determining if selection for plasticity under somatic hypermutation has operated in Ig genes, we turn the question around. We ask whether evolutionary evidence in the form of site-specific codon bias supports a role for somatic hypermutation in TCRs. The answer is: probably. As we will show, neither murine V_γ nor human V_γ nor human $V_{\alpha/\delta}$ genes show any indication of selection for site-specific codon bias, but both human and murine V_β and, to a lesser extent, murine $V_{\alpha/\delta}$ genes show codon mutability patterns similar to, but weaker than, those seen in Igs. This result is consistent with the idea that some TCRs hypermutate in a functionally meaningful way, or have done so in the evolutionary past, and that selection has operated to enhance their plasticity under affinity maturation in much the same way as has happened for Igs (KEPLER 1997).

2 Methodology

One of us (TBK) has previously performed a statistical analysis of site-specific codon bias and mutability in immunoglobulins using a pair of databases. The first, used to establish the sequence-specificity of the hypermutation, contains 520 unselected mutations (out of 28 511 bases sequenced) in the J_H-C_H and J_κ-C_κ introns and was published by SMITH et al. (1996) and will be referred to here as the intron database. The second database, which we used to determine the site-specific codon bias of human Ig V region genes, is the VBASE Sequence Directory of human germline Ig V genes, collected and maintained by TOMLINSON et al. (1996). Here, we will also analyze the databases of human and murine TCR V region genes published by ARDEN et al. (1995a, b). For all V genes, only genes that appeared to be functional were included in the analyses.

In addition to the V gene databases, we assembled two more databases as controls. One is an alignment of 11 human Ig constant-region genes (first Ig-domain). The other is a collection of 11 human genes that belong to the Ig superfamily but are not suspected of being subject to somatic hypermutation (for further details, see KEPLER 1997).

2.1 Example: GAC vs GAT

The method employed is most easily illustrated by focusing attention at first on the synonymous codons for a single amino acid, for example, aspartic acid, whose codons are GAC and GAT. We turn first to the intron database to evaluate the relative mutability of these two nucleotide triplets. In the germlines corresponding to each of the sequences in the intron database, there are 438 occurrences of GAT. Among these occurrences are 33 cases where one of the three nucleotides has mutated in such a way that the triplet, if it were in frame, would encode something other than aspartic acid. Therefore, we define a mutability for GAT as $f_{GAT} = 33/(3 \times 438) = 2.51\%$. In contrast, there are 463 occurrences of GAC in the intron database with just four replacement mutations, for a GAC mutability of $f_{GAT} = 33/(3 \times 438) = 0.29\%$.

Having computed the mutabilities, we turn to the germline V gene databases. Among the functional human V_H genes in VBASE there are 137 appearances of the codon GAT, 104 of which are in CDRs as defined by KABAT et al. (1991). We define a CDR localization index for GAT: $F_{GAT} = 104/137 = 0.759$. For GAC we have 102 CDR occurrences out of 488 altogether, so that $F_{GAC} = 102/488 = 0.209$. For this amino acid, the more mutable codon is more localized in to CDRs than the less mutable one.

2.2 Mutability and CDR Localization for all Codons

In order to extend this comparison to all amino acids and all informative codons, we want to subtract out the confounding effects of selection for the amino acids themselves to isolate selection for the codons. So for each codon XYZ, we compute the mean mutability over all codons that encode the same amino acid as does XYZ. This mean value is then subtracted from the mutability of XYZ to yield its differential mutability, i.e., its mutability relative to the set of its synonymous codons. In this way, we get a differential mutability for GAT of $\delta_{GAT} = f_{GAT-1/2}(f_{GAT} + f_{GAC}) = +2.2\%$. Similarly, we have, for GAC, $\delta_{GAC} = -2.2\%$. We likewise subtract the mean CDR localization to obtain the differential localization, $\Delta_{GAT} = F_{GAT-1/2}(F_{GAT} + F_{GAC}) = +0.275$ and similarly, $\Delta_{GAC} = -0.275$. These indices are computed for all informative codons in exactly this way.

2.3 Linear Correlation Coefficient

If V genes have been selected for plasticity under somatic mutation, the differential mutability of triplets will be positively correlated with the differential localization of triplets into the CDRs. The appropriate statistic to examine, then, is the linear correlation coefficient, r (Steel et al. 1996).

The expected value of r is just zero if the relative mutability and relative CDR localization are independent, but is positive if they are correlated. To determine how large r must be to indicate more than just chance fluctuations, we need to compute the number of degrees of freedom. The number of degrees of freedom is given by the number of measurements minus the number of constraints and estimated parameters. In this case that number is 38: 38 = 64 codons–3 stop codons–2 uninformative codons (methionine and tryptophan are uniquely encoded)–2 codons absent from the intron database–18 estimated average mutabilities (one for each informative amino acid)–1 estimated regression coefficient. Now we can use standard techniques to evaluate the significance of our numbers, or the probability of getting the results we do under the null hypothesis that the mutability and CDR localization are independent (Steel et al. 1996).

2.4 Difference Between Mean Mutability in CDRs and FRs

We also tested a slightly different hypothesis: that the average differential mutability of codons at CDR positions differs from the average differential mutability of codons in FRs. We computed the mean differential mutability over all appropriate genes at each position in the alignment, and then averaged separately over all CDR positions and over all FR positions. For this difference we performed a t-test. The mean mutability difference test and the linear correlation test are not entirely independent, but do provide, as we will see, complementary information.

2.5 The Problem with Serine

Previous evidence for the targeting of mutation into CDRs was presented by Wagner et al. (1995) based on the codon-usage bias of serine. This was an important advance, but the analysis was flawed. Multiple occurrences of serine codons at the same position in different related genes are not independent and the binomial test used in their study (which assumes that they are independent) underestimates the p value sometimes quite substantially. The problem is exacerbated by the unique feature of serine among amino acids of being coded by two disjoint sets of codons.

Serine codons fall into two classes that cannot be mutated one to another in a step-wise manner without passing through a non-serine residue in an intermediate stage. The serine codons also show the greatest range in mutability of all codons. This is unfortunate for us because maintenance of a mutable serine codon at a given position in an alignment could mean either that mutability itself has been selected for, or that serine has been strongly conserved and it has simply not been possible to move from a more mutable codon (AGY) to a less mutable codon (CTN). Serine is the only amino acid for which this is true.

Therefore, we have done an additional pair of tests, identical to those described above but from which all serine codons are excluded.

3 Results

3.1 Immunoglobulins: Human V_H, V_λ and V_κ

The results of our tests are summarized in Table 1. The upper rows of Table 1 display our results for human Ig V genes. The effect is most pronounced in V_H genes where the findings are unambiguous. Both the linear correlation coefficient, r, and the t statistic for the CDR-FR differential mutability are very significantly positive whether or not serine codons are included (p values of 0.004 and 0.003 for r and t, respectively, when serine codons are included, and 0.012 and 0.003 when serine codons are excluded). The results for Ig V should also be contrasted with those for the "control" alignments at the bottom of Table 1.

More compelling even than these statistics is the plot of mutability against amino acid position averaged over the V_H alignment (Fig. 1). In this figure it is striking that the differential mutabilities in FR1 and FR2 are so uniformly negative. Note also that the average mutability over the Ig superfamily control database is very nearly zero: $-0.062 \pm 0.004\%$ (Kepler 1997), so that the negative values seen in the figures shown here really do represent values more negative than background.

The V_λ sequences also yield significant results when serine codons are included, but when serine codons are excluded, the statistics become much less compelling. Closer inspection of the alignments and mutability by position (Fig. 2) shows that a tandem pair of serine codons within CDR3 and serine codons within CDR1 of many V_λ sequences contribute strongly to both the r and t statistics. When serine codons are excluded, the correlation drops (but remains somewhat high) and the t-test falls precipitously. This does not mean that there is no evidence for

Table 1. The correlation coefficient *r* and *t*-statistic for several gene alignments

Locus	All triplets		No serine	
	r	*t*	*r*	*t*
hV_H	+0.446*	+3.19*	+0.419*	+3.24*
hV_κ	+0.422*	+3.48*	+0.291**	+0.61
$h\ V_\lambda$	+0.219	+1.53	+0.187	+1.26
hV_β	+0.295**	+0.76	+0.471*	+1.54
hV_βa	+0.540*	+3.41*	+0.479*	+1.83**
$hV_{\alpha/\delta}$	−0.035	−0.14	+0.075	+0.30
hV_γ	−0.164	−1.11	−0.063	−0.43
mV_β	+0.295**	+0.972	+0.376*	+1.72**
mV_βa	+0.531*	+3.62*	+0.377*	+2.02*
$mV_{\alpha/\delta}$	+0.172	+1.70**	+0.345*	+3.21*
mV_γ	+0.238	−1.71**	−0.072	−1.74**
Ig C	+0.157	−0.75	+0.156	−0.47
Ig SF	−0.012	−0.47	−0.155	−0.56

*These numbers are significant at the 0.05 level or higher (in a two-tailed test).
**These numbers are significant at the 0.1 level, and are included for comparison only.
[a]These V_β entries have been analyzed using a non-standard definition of CDR3 in which the first CDR3 amino acid is 94 (Kabat numbering) rather then 95 (see discussion in text). Ig C is an alignment of human constant-region genes (first Ig domain) used here as a control, and Ig SF is an unaligned collection of immunoglobulin superfamily genes from human, again as a control (see Kepler, 1997).
The *h* or *m* before the gene locus name indicates human or mouse, respectively (all Igs alignments are human).

site-specific codon bias in V_λ; excluding serine codons means that we ignore what might well be the largest part of the effect we seek to document. But it does mean that additional evidence should be sought. This evidence is available by visual inspection of the position-by-position mutability. In fact, most sites in V_λ have lower than average mean mutability, in both CDRs and FRs, suggesting selection

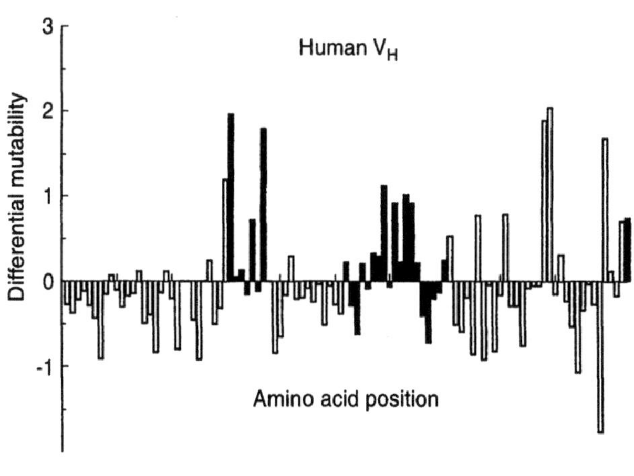

Fig. 1. The average differential mutability (as a percentage) at each amino acid position for the V_H alignment. The *open bars* designate positions within FRs and the *solid bars*, those within CDRs. [Modified from Kepler (1997)]

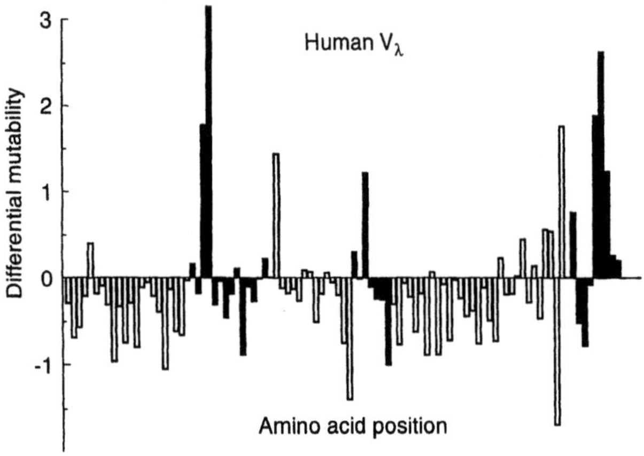

Fig. 2. The average differential mutability (as a percentage) at each amino acid position for the V_λ alignment. The *open bars* designate positions within FRs and the *solid bars*, those within CDRs. [Modified from Kepler (1997)]

for lower mutability across most of the gene segment. It seems that for V_λ, the effect of codon bias is to lower the mutability overall.

The numbers for V_κ by themselves show very little evidence for site-specific codon bias. Figure 2, however, does reveal an interesting pattern. CDR1 is quite clearly a zone of mutability, while FR1 and FR2 are uniformly less mutable than average. But CDR2, FR3 and CDR3 all show little effect. More dramatically, there is a single pair of codons in FR3 that increase the mean mutability of FR3 (Fig. 3). These turn out to be serine codons again, in most genes of the alignment. Corresponding codons can be found in V_H as well, where a localized increase of mutability in FR3 is seen (see Fig. 1). We cannot determine at this time whether the mutability of these codons is important in their conservation or if they are simply frozen-in by the serine effect mentioned above. Nevertheless, the visual pattern revealed in Fig. 1 strongly suggests that V_κ has also experienced selection for plasticity under somatic hypermutation.

The patterns of mutability differ remarkably from one V gene locus to another. The greatest mutability in V_H is found in CDR2 (see Fig. 1), whereas that for V_λ is in CDR3 (Fig. 2), while V_κ is clearly most mutable in CDR1 (see Fig. 3). Whether the differences in local structure of the mutability patterns between genes of different loci are indicative of underlying functional differences between these genes we cannot at this time say. Similarly, it is an interesting open question why FR3 shows greater variability with regard to its mutability pattern than either FR1 or FR2 in all Ig V genes.

3.1.1 Amino Acid Usage

Mutability may also be affected by amino acid usage in CDRs and FRs. However, since different amino acids will be under different selective constraints in CDRs and

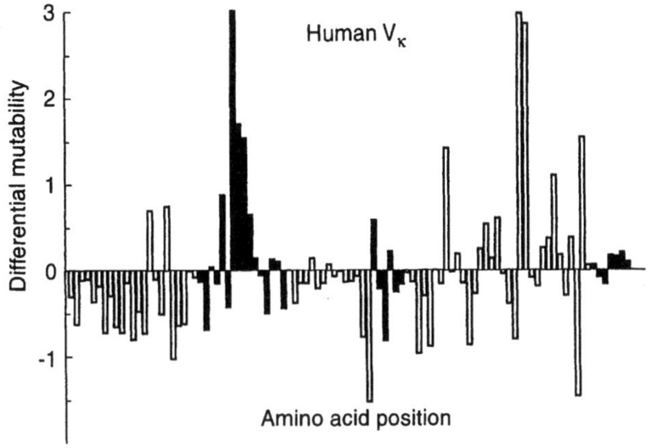

Fig. 3. The average differential mutability (as a percentage) at each amino acid position for the V_κ alignment. The *open bars* designate positions within FRs and the *solid bars*, those within CDRs. [Modified from KEPLER (1997)]

FRs, we cannot statistically isolate the effects of selection for mutability on amino acid usage. Nevertheless, it is of some interest to compare the total mutability of CDRs and FRs, including the effects of amino acid usage. We find that in V_H, the CDRs are 48% more mutable overall than FRs with 67% of this difference accounted for by codon bias. In V_λ and V_κ, the total mutability of CDR is 48% and 26% greater than the total mutability of FR; codon bias accounts for 87% and 69% of this difference, respectively. In all cases, differential amino acid usage acts in concert with codon bias to enhance the mutability differences.

3.2 TCRs: Human and Murine V_β, $V_{\alpha/\delta}$ and V_γ

We use the same mutability data obtained from the intron mutations database that we applied to the Ig V genes and apply it, in precisely the same way, to alignments of TCR V genes from both mice and humans (ARDEN et al. 1995a, b).

We use the definitions of CDRs and FRs established by KABAT et al. (1991), although there is more uncertainty in their true placement for TCR than for Ig. It may appear odd to analyze TCRs by considering all three CDRs in the same way we do for Igs when, in fact, correlative studies of TCR structure and antigen-MHC specificities indicate that the CDR1s and CDR2s interact with MHC molecules while only the CDR3s seem to have a prominent influence on antigenic peptide interactions (see, e.g., FINK et al. 1986; ACHA-ORBEA et al. 1988; ENGEL et al. 1988; SORGER et al. 1990). Various models have proposed that the CDR3s contact the peptide while the CDR1s and the CDR2s bind the α-helices of the MHC molecule (CHOTHIA et al. 1988; DAVIS and BJORKMAN 1988; CLAVERIE et al. 1989; JORGENSEN et al. 1992). On the other hand, recent crystallographic evidence indicates that, for at least one TCR, all three CDRs are positioned over the peptide to some extent (GARCIA et al. 1996; see also the review by BENTLEY and MARIUZZA 1996).

Therefore, mutations in any one of the CDRs have the potential to effect the interaction of the TCR with the antigenic peptide.

The results we obtain are somewhat surprising, even if not unambiguous. Both human and murine V_β show a positive trend, particularly in the correlation coefficient. Closer inspection of murine V_β (Fig. 4) again shows that there is a hotspot consisting of two codons, at Kabat positions 94 and 95. CDR3 is thought to begin at Kabat position 95, but the residue at position 94 could also play a role in antigen binding and therefore be subject to positive selection for mutability. When we assign position 94 to CDR3 rather than to FR3 (see Table 1, rows corresponding to hV_β and mV_β), both r and t become significant. Again, the hot positions are, to a large extent but not exclusively, due to serine codons. Excluding serine codons from the analysis, however, still leaves a significant result for r and a borderline result for t both human and murine V_β genes. In other words, the result is not entirely an artifact of the serine effect. Whereas the ambiguity in the statistics for the case of V_κ was satisfactorily resolved by more closely inspecting the average mutability position-by-position (see Fig. 2), the corresponding plot for V_β (see Fig. 3) is frustratingly inconclusive. While CDR2 does appear to be a zone of excess mutability (and, of course, the very high mutability of the codons at positions 94 and 95 are apparent here), the FRs are much less uniform than the FRs in the Ig V genes (Figs. 1–3) and there is less variability overall in V_β.

The V_γ genes show no trend. Note that the borderline result in the t statistic for murine V_γ is not particularly meaningful. The t value is negative, indicating a slightly higher mutability in the FRs than in the CDRs, opposite to what one expects under the hypothesis we are proposing, but the correlation coefficients, r, are positive. So the two tests do not corroborate each other, and neither of them is significant alone. This also illustrates the lack of complete dependence of one test on the other. That is, the t-test does give information not contained in the correlation test.

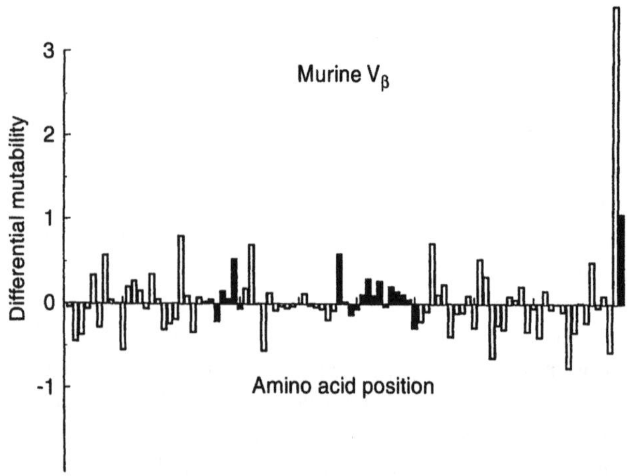

Fig. 4. The average differential mutability (as a percentage) at each amino acid position for the V_β alignment. The *open bars* designate positions within FRs and the *solid bars*, those within CDRs

The human $V_{\alpha/\delta}$ genes show no evidence for the effect we seek, but the murine $V_{\alpha/\delta}$ genes show a significant effect when serine codons are excluded. This result is not as puzzling as it may seem. Just as the serine effect could "freeze in" the appearance of plasticity enhancement via codon bias, it could also "freeze out" the effect even if the codons for all other amino acids have been affected. Whether or not this is the case for murine $V_{\alpha/\delta}$ genes we cannot at this point say.

4 Discussion

It is quite unlikely that the distinctive pattern of site-specific codon bias in human Ig V genes has arisen by chance alone. It is more likely that these patterns resulted from an advantage gained from the use of more mutable codons in the CDRs and less mutable codons in the FRs. This bias enhances the plasticity of these molecules under somatic mutation and consequently increases the efficiency of the germinal center reaction. The logic is clear, but it remains surprising that the advantage gained over evolutionary time is sufficient to produce a recognizable signal in extant genes, especially in light of the enormous turnover (and presumed low cost) of B cells. This apparent disposability of B cells may be somewhat misleading in this context, however. The cells that undergo somatic mutation have already proven their worth by binding to a real and present antigen. The selective advantage may be due to the enhanced retention of these more valuable B cell specificities during affinity maturation.

Furthermore, the selective advantage of plasticity under somatic mutation increases sharply with the overall somatic mutation rate. Where this rate is extremely high, the advantage of directing mutability can be quite considerable, if not essential, because the loss of previously selected antigen receptors to deleterious mutations becomes quite severe (Kepler and Perelson 1993). An alternative strategy to dealing with the difficulties of a very large somatic mutation rate is to simply lower the rate. That codon-bias induced plasticity has evolved instead speaks strongly for the importance of affinity maturation in the overall strategy of effective immunity.

We know that somatic mutation of Ig genes occurs and we asked if this process has left an evolutionary trace. Having found that it does, we inverted the analysis applied to Igs for application to TCRs. This approach is a novel way of addressing the controversial question of somatic mutation in TCRs. We sought additional evidence by looking for site-specificity in TCR V genes of the same type that we found in Ig V genes. The results are not definitive, by any means; the evolutionary traces are not as strong for the TCRs as they are for the Igs, but they do add weight to the claims that TCRs hypermutate and further suggest that TCR hypermutation is, or was, functionally significant.

What we have done represents only a beginning. Much more sophisticated analyses need to be undertaken. Our present analyses are hampered by several problems. The statistical tests used, while adequate for clear cases like that of human V_H, are nevertheless weakened by the lack of a robust model for the underlying evolutionary processes. Improvement in our statistical methods would naturally be tied to a greater understanding of the evolutionary dynamics of large gene families. Compounding our problems is the role of serine codons. The combination of their prominence as targets of somatic hypermutation and the uniqueness of their genetic encoding (and consequently the uniqueness of their evolutionary dynamics) demands greater attention than we have been able to provide. Another difficulty is that the database of unselected mutations used to quantitate the mutability of trinucleotides, though large by ordinary standards, is actually somewhat small for the present purposes. As more unselected data from a larger variety of genes becomes available, the resolution of our methods should increase concomitantly and examination of finer details of Ig and TCR structure–function relationships through exploration of the traces left by their evolutionary history will become possible. Lastly, our analyses have focused on the mutability of codons independent of the context in which they are found, i.e., independent of flanking DNA sequence information. Since the mutation mechanism (presumably) recognizes sequence motifs without regard to reading frame, additional variability will be attributed to neighboring nucleotides. Once these methodological difficulties have been adequately addressed, we expect to apply them across phylogenies and to a larger variety of genes to gain further insight into the role of somatic hypermutation in the evolution of lymphocyte antigen receptors.

Acknowledgements. This work was supported by grant number MCB-9357637 from the National Science Foundation (TBK).

References

Acha-Orbea H, Mitchell DJ, Timmermann L, Wraith DC, Tausch GS, Waldor MK, Zamvil SS, McDevitt HO, Steinman L (1988) Limited heterogeneity of T cell receptors from lymphocytes mediating autoimmune encephalomyelitis allows specific immune intervention. Cell 54:263–273

Arden B, Clark SP, Kabelitz D, Mak TW (1995a) Human T-cell receptor variable gene segment families. Immunogenetics 42:455–500

Arden B, Clark SP, Kabelitz D, Mak TW (1995b) Mouse T-cell receptor variable gene segment families. Immunogenetics 42:501–530

Bachl J, Wabl M (1995) Do T-cells hypermutate? Nature 375:286

Bentley GA, Mariuzza RA (1996) The structure of the T cell antigen receptor. Annu Rev Immunol 14:563–590

Betz AG, Rada C, Pannell R, Milstein C, Neuberger MS (1993) Passenger transgenes reveal intrinsic specificity of the antibody hypermutation mechanism: clustering, polarity and specific hot spots. Proc Natl Acad Sci USA 90:2385–2388

Chang B, Casali P (1994) The CDR1 sequences of a major proportion of human germline Ig V_H sequences are inherently susceptible to amino acid replacement. Immunol Today 15:367–373

Chothia C, Boswell DR, Lesk AM (1988) The outline structure of the T-cell αβ receptor. EMBO J 7(12):3745–3755

Claverie J-M, Prochnicka-Chalufour A, Bouguelerat L (1989) Implications of a Fab-like structure for the T-cell receptor. Immunol Today 10:10–14

Davis MM, Bjorkman PJ (1988) T-cell antigen receptor genes and T-cell recognition. Nature 334:395–402
Dénepoux S, Razanajoana D, Blanchard D, Meffre G, Capra JD, Banchereau J, Lebecque S (1997) Induction of somatic mutation in a human B cell line in vitro. Immunity 6:35–46
Du Pasquier L (1993) Evolution of the immune system. In Paul W (1993) Fundamental Immunology, 3rd edn. Raven, New York, pp 199–234
Engel I, Hedrick SM (1988) Site-directed mutations in the VDJ junctional region of a T cell receptor beta chain cause changes in antigenic peptide recognition. Cell 54:473–484
Fink PJ, Matis LA, McElligott DL, Bookman M, Hedrick SM (1986) Correlations between T cell specificity and the structure of the antigen receptor. Nature 321:219–226
Garcia KC, Degano M, Stanfield RL, Brunmark A, Jackson MR, Peterson PA, Teyton L, Wilson IA (1996) An alpha beta T cell receptor structure at 2.5 A and its orientation in the TCR-MHC complex. Science 274:209–219
Grantham R, Gautier C, Gouy M (1980) Codon frequencies in 119 individual genes confirm consistent choices of degenerate bases according to genome type. Nucleic Acids Res 8:1893–1912
Hinds-Frey KR, Nishikata H, Litman RT, Litman GW (1993) Somatic variation precedes extensive diversification of germline sequences and combinatorial joining in the evolution of immunoglobulin heavy chain diversity. J Exp Med 178:815–824
Hughes AL, Nei M (1988) Pattern of nucleotide substitution at major histocompatibility complex class I loci reveals overdominant selection. Nature 335:167–170
Jorgensen JL, Reay PA, Ehrich EW, Davis MM (1992) Molecular components of T-cell recognition. Annu Rev Immunol 10:835–73
Kabat EA, Wu TT, Perry HM, Gottesman KS, Foeller G (1991) Sequences of proteins of immunological interest, 5th edn. Department of Health Services, National Institutes of Health, Bethesda, pp 2130–2146
Kelsoe G (1996) Life and death in germinal centers (redux). Immunity 4:107–110
Kelsoe G, Zheng B, Kepler TB (1995) Do T-cells hypermutate? Nature 375:286 (response to Bachl and Wabl)
Kepler TB (1997) Codon bias and plasticity in immunoglobulins. Mol Biol Evol 14:637–643
Kepler TB, Perelson AS (1993) Somatic hypermutation in B cells: an optimal control treatment. J Theor Biol 164:37–64
Lewis SM (1994) The mechanism of V(D)J joining: lessons from molecular, immunological, and comparative analyses. Adv Immunol 56:27–150
Maynard-Smith J, Smith NH (1996) Site-specific codon bias in bacteria. Genetics 142:1037–1043
Möller G (1990) (ed) Immunol Rev 115
Motoyama N, Okada H, Azuma T (1991) Somatic mutation in constant regions of mouse lambda_1 light chains. Proc Natl Acad Sci 88:7933–7373
Reynaud CA, Garcia C, Weill J-C (1995) Hypermutation generating the sheep immunoglobulin repertoire is an antigen-independent process. Cell 80:115–125
Rogozin IB, Kolchanov NA (1992) Somatic hypermutagenesis in immunoglobulin genes. II. Influence of neighboring base sequences on mutagenesis. Biochim et Biophys Acta 1171:11–18
Rothenfluh H, Blanden RV, Steele EJ (1995) Evolution of V genes: DNA sequence structure of functional germline genes and pseudogenes. Immunogenetics 42:159–171
Smith AS, Creadon G, Jena PK, Portanova JP, Kotzin BL, Wysocki LJ (1996) Di- and trinucleotide target preferences of somatic mutagenesis in normal and autoreactive B cells. J Immunol 156:2642–2652
Sorger SB, Paterson Y, Fink PJ, Hedrick SM (1990) T cell receptor junctional regions and the MHC molecule affect the recognition of antigenic peptides by T cell clones. J Immunol 144(3):1127–1135
Steel RGD, Torrie JH, Dickie DA (1996) Principles and procedures of statistics: a biometrical approach, 3rd edn. McGraw-Hill, New York, pp 286–299
Stewart J, Coutinho A (1996) (eds) The evolutionary origins of immunoglobulins and T-cell receptors. Res Immunol 147
Tanaka T, Nei M (1989) Positive Darwinian selection observed at the variable-region genes of immunoglobulins. Mol Biol Evol 6:447–459
Tomlinson IM, Williams SC, Corbett SJ, Cox JBL, Winter G (1996) V BASE sequence directory MRC centre for protein engineering. Cambridge, UK (URL: http://www.mrc-cpe.cam.uk/imt-doc/vbase-home-page.html)
Van der Stoep N, Van der Linden J, Logtenberg T (1993) Molecular evolution of the human immunoglobulin E response: high incidence of shared mutations and clonal relatedness among V_H5 transcripts from three unrelated patients with atopic dermatitis. J Exp Med 177:99–107

Varade WS, Marin E, Kittelberger AM, Insel RA (1993) Use of the most J_H-proximal human Ig H chain V region gene, V_H6, in the expressed immune repertoire. J Immunol 150:4985–4995
Wagner SJ, Neuberger MS (1996) Somatic hypermutation of immunoglobulin genes. Annu Rev Immunol 14:441–458
Wagner SJ, Milstein C, Neuberger MS (1995) Codon bias targets mutation. Nature 376:732
Wilson M, Hsu E, Marcuz A, Courtet L, Du Pasquier L, Steinberg C (1992) What limits affinity maturation of antibodies in Xenopus – the rate of somatic mutation or the ability to select mutants? EMBO J 11:4337–4347
Zheng B, Xue W, Kelsoe G (1994) Locus-specific somatic hypermutation in germinal centre T cells. Nature 372:556–559

Theoretical Limits to Massive Receptor Editing in Immature B Cells

D. NEMAZEE

1 Introduction

Secondary immunoglobulin light chain gene rearrangements occur frequently during B cell development (LEWIS et al. 1982; VAN NESS et al. 1982; FEDDERSEN and VAN NESS 1985; SHAPIRO and WEIGERT 1987; CLARKE and MCCRAY 1991; HARADA and YAMAGISHI 1991). This process can provide the B cell numerous attempts at generating functional antigen receptors. On the κ locus such secondary rearrangements can remove and replace assembled VJ elements from their context adjacent to the constant region. This type of rearrangement is possible because the gene organization of this locus provides the continual presence of unrearranged V regions upstream of joined V/J coding segments. Through the normal process of V-J recombination, these V gene segments can in turn join to previously unused J segments downstream of the joined V/J coding segment, yielding nested, secondary rearrangements (TONEGAWA 1983). In addition, in λ-bearing cells recombining sequence (RS) rearrangements often delete the C_κ loci (DURDIK et al. 1984; MOORE et al. 1985).

There is increasing evidence that cells bearing in-frame light chain gene rearrangements do not always, and perhaps only rarely, stop further rearrangements (KWAN et al. 1981; FEDDERSEN and VAN NESS 1985; HARDY et al. 1986; GOLLAGHON et al. 1988; LEVY et al. 1989; HUBER et al. 1992; MA et al. 1992; GAY etal. 1993; RADIC et al. 1993; TIEGS et al. 1993; CHEN et al. 1994; DOGLIO et al. 1994; PRAK et al. 1994, 1995; VERKOCZY et al. 1995). The potential explanations for such continuing nested rearrangements are many. Some light chains may be non-functional because of an inability to assemble with the cell's μ-heavy chain (KWAN et al.

Division of Basic Sciences, Department of Pediatrics, National Jewish Center for Immunology and Respiratory Medicine, 1400 Jackson Street, Denver, Colorado 80206, USA

1981). Another possibility is that weak light chain protein expression leads to the elimination of a functional light chain before the gene's functionality is "apparent" to the cell (DOGLIO et al. 1994). It has been proposed that self antigen-mediated "positive selection" may be required for B cell maturation – in analogy to the critical step in T cell development (MA et al. 1992). Finally, self antigen interacting with the surface immunoglobulin may promote ongoing rearrangements (GAY et al. 1993; RADIC et al. 1993; TIEGS et al. 1993; CHEN et al. 1994; PRAK et al. 1994). This latter process of antigen-mediated ongoing rearrangement is referred to as receptor editing.

Data supporting a putative critical role of receptor editing in self tolerance is derived from a number of studies in transgenic mice bearing autoreactive B cell specificities (GAY et al. 1993; RADIC et al. 1993; TIEGS et al. 1993; CHEN et al. 1994; PRAK et al. 1994), and in κgene targeted mice expressing germline functional κ genes in the normal context of the κ locus (PRAK et al. 1995). Immature B cells encountering autoantigen overexpress recombinase activator gene (RAG) mRNA and undergo secondary κ and λ rearrangements (TIEGS et al. 1993). An antigen expressed exclusively in the periphery failed to induce receptor editing, suggesting that tolerance through receptor editing is restricted to the immature stage of B cell development (RUSSELL et al. 1991; TIEGS et al. 1993). B cells expressing heavy chains with a propensity to give rise to autoantibodies are able to very efficiently select light chains that avoid autoreactivity through multiple light chain gene rearrangement attempts (GAY et al. 1993; RADIC et al. 1993; CHEN et al. 1994; PRAK et al. 1994). In about 80% of cells examined, cells bearing targeted, in-frame κ genes underwent subsequent light chain rearrangement (PRAK et al. 1995). In this case it is difficult to argue that the context or tissue specific regulation of the light chain gene in question was responsible for this poor allelic exclusion. Rather, it would appear quite likely that receptor editing is a normal genetic mechanism that occurs in many B cells. Indeed, isolation of circular DNA excision products from normal B cells has identified frequent deletion of in-frame VJ_{κ} rearrangements by secondary rearrangements (HARADA and YAMAGISHI 1991). Furthermore, we have argued elsewhere that autoreactivity in the unselected B cell repertoire should be extremely frequent, and that roughly 63% of all cells should be deleted through tolerance (NEMAZEE 1996). This high frequency of autoreactivity would in turn predict that receptor editing mechanisms would be very useful in providing a way to salvage cells that would otherwise be eliminated.

Two sets of data in particular present a paradox: the work of HARADA and YAMAGISHI (1991) and PRAK et al. (1995) has suggested that in-frame light chains only rarely prevent secondary light chain rearrangements, whereas the high frequency (approximately 40%–60%) of κ^+ B cells bearing one κ locus in the germline configuration (κ^+/κ^o) intuitively suggests that secondary rearrangements must be infrequent (COLECLOUGH et al. 1981; ZOU et al. 1993). In the present study we attempt to calculate the theoretical constraints limiting the frequency of occurrence of secondary light chain rearrangement, and to determine the potential contributions of self tolerance and light chain "non-functionality" to these secondary rearrangements.

2 Elements of Theoretical Framework

It is well accepted that two thirds of light chain rearrangements are out-of-frame and that these rearrangements fail to mediate allelic exclusion. The present model attempts to account for the apparently large numbers of cells that continue to rearrange light chain genes despite the presence of in-frame rearrangements. Among the in-frame rearrangements, some light chains may fail to be expressed, others may be pseudogenes that bear stop codons or mutations that disrupt structure. Still other light chains may be unable to pair with the cell's heavy chain because of idiosyncratic structural incompatibilities, i.e., they may be able to pair with many, but not all heavy chains. As is the case with out-of-frame rearrangements, all defects of in-frame light chains leading to non-functionality should be *recessive* with respect to allelic exclusion, that is, the presence of a single non-functional light chain chromosome should fail to block the feedback suppression of rearrangement induced by a second, functional light chain. On the other hand, light chains that confer autoreactivity should be *dominant* with respect to the induction of secondary rearrangements, and must be inactivated by receptor editing in order for recombination to be suppressed by a second light chain allele that is both functional and non-autoreactive. (A final element that will not be explicitly considered here is the notion of a requirement for positive selection based on ligand binding for the downregulation of recombination. Such positive selection predicts that light chains that allow surface IgM expression, but fail to encounter positively selecting ligands, should be recessive with respect to the mediation of allelic exclusion.)

With these elements in mind, one can introduce two variables: a, the probability that an in-frame rearrangement is non-autoreactive, and f, the probability that an in-frame rearrangement is "functional". Thus the values 1-a and 1-f represent the likelihoods that in-frame κ rearrangements fail to turn off V_κ–J_κ recombination because of dominant or recessive factors, respectively.

Let us first consider the simplest case in which $f = 1$, i.e., all in-frame rearrangements are able to make functional HL pairs. Consider an autoreactive bone marrow B cell that has a κ light chain configuration of SR/–, where SR refers to the chromosome encoding the functional light chain that contributes to the self-reactive specificity and "-" refers to a non-functional or germline configuration on the other allele. The probability that a single, secondary κ rearrangement will silence and replace the SR allele, resulting in a +/– cell, is $1/2 \cdot 1/3\ a = 1/6a$ (where a is the probability that the newly generated specificity is not self reactive). The majority of rearrangement attempts will either fail to alter the self specificity (yielding cells of type SR/–, SR/+, SR/SR) or result in two out-of-frame alleles (–/–). However, subsequent attempts could result in rescue of these non-functional rearrangements at frequencies that are dependent on the particular rearrangement status of the cells. (For example, the fractional rescue rates for initial rearrangement classes SR/–, SR/+, SR/SR in a subsequent rearrangement are given as $1/6a$, $1/3 + 1/6a$, and zero, respectively.) Assuming that multiple attempts at secondary rearrangements are possible, one can then assess the cumulative recovery of functional, non-autoreactive cells with respect to the number of rearrangement attempts. Figure 1 provides some cumulative salvage rates as a function of *a* and the number of rearrangement attempts. As can be seen, receptor editing can efficiently rescue autoreactive cells, even when the probability of autoreactivity is high.

Incorporating the notion of light chain non-functionality, it can be seen that, in cells with no previous rearrangements, and therefore two non-functional κ alleles (−/−), the frequency of generation of non-functional rearrangements per attempt is 1–1/3*f*, which includes also out-of-frame joins (Fig. 2). Among the 1/3*f* functional rearrangements, the proportion that are non-autoreactive, and therefore mediate allelic exclusion, is *a*, thus the overall frequency of cells that stop further rearrangements (category +/−) is 1/3*fa*. The remaining fraction, 1/3*f*–1/3*fa*, is the proportion of cells that will undergo receptor editing, i.e., autoantigen-induced secondary rearrangements (category SR/−). The proportions of successful rearrangement attempts is dependent on the starting status of the cell. For example, in autoreactive cells (SR/−) the proportion of +/− cells generated in a single attempt is 1/6*fa*, which is half the proportion obtained with −/− starting cells, the remaining 50% of rearrangements generate SR/+ cells, which are unsuccessful at mediating allelic exclusion and developmental progression. Overall, there are six possible states of the cell and the probabilities of entering different states or classes in a single rearrangement attempt defines a six-by-six matrix (Table 1). In the present model it is assumed that functional, non-autoreactive rearrangements are 100% efficient at mediating allelic exclusion. Despite this perfect efficiency in mediating

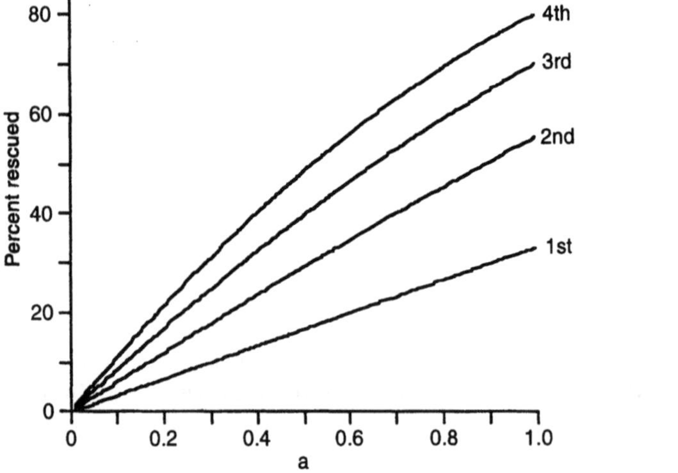

Fig. 1. Cumulative rescue of autoreactive cells through secondary rearrangement attempts as a function of *a*, the fraction of B cells that are non-autoreactive. Please note the substantial rescue of autoreactive cells that is possible even when autoreactivity is frequent (i.e., when *a* is low)

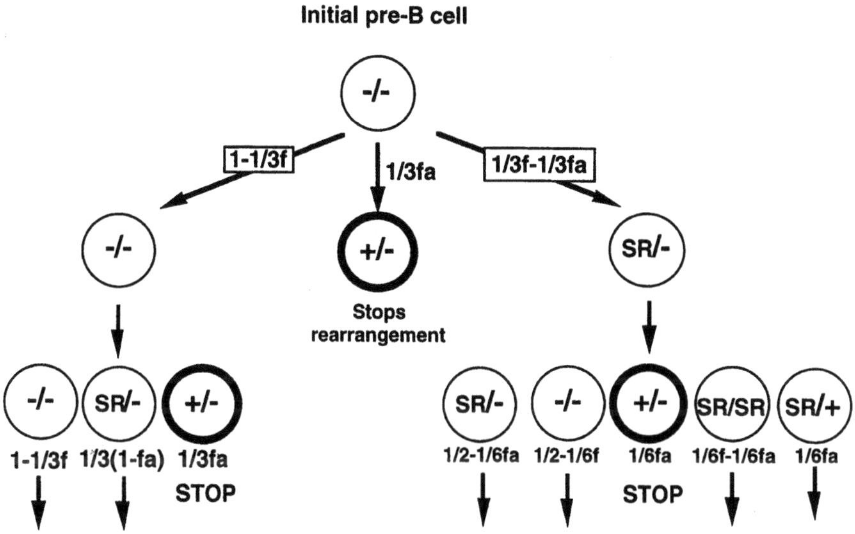

Fig. 2. A model for light chain allelic exclusion that takes into account light chain non-functionality and autoreactivity-induced receptor editing. Fraction of functional cells generated: first round, 1/3fa; second round, $(1-1/3f)(1/3fa) + (1/3f-1/3fa)(1/6fa)$; third round, etc.

Table 1. Branching ratios defining the probability of conversion from one rearrangement class to another after a single rearrangement. Depletion of Js is not taken into account in this scheme and therefore it is only valid for early rounds of rearrangement attempts.

Initial category	Subsequent category					
	−/−	+/−	+/+	−/SR	+/SR	SR/SR
−/−*	$1-1/3f$	$1/3fa$	0	$1/3f-1/3fa$	0	0
+/−	0	1	0	0	0	0
+/+	0	0	1	0	0	0
−/SR	$1/2-1/6f$	$1/6fa$	0	$1/2-1/6fa$	$1/6fa$	$1/6f-1/6fa$
+/SR	0	$1/2-1/6f$	$1/6fa$	$1/2-1/6f$	$1/6f$	$1/6f-1/6fa$
SR/SR	0	0	0	$1-1/3f$	1/3fa	1/3f-1/3fa

* Indicates unrearranged or out of frame.

allelic exclusion, "doubles", i.e., cells with two functional, non-autoreactive rearrangements (+/+) can be generated in this scheme from SR/+ intermediates. We have assumed for simplicity that SR/+ cells never block further rearrangements, although such cells may sometimes lose self reactivity if the H chain is limiting and pairs preferentially with the + light chain.

Critical variables are the maximal allowed V_κ–J_κ rearrangements/cell, which defines particular ranges of mean V_κ–J_κ rearrangements/cell. The maximal allowed rearrangement attempts is related to the so-called "crash factor" (LANGMAN and COHN 1995), which limits the effective time window in which B cells must generate functional receptors, whereas the mean rearrangements/cell is the measure of the average rearrangements/cell obtained in the population.

Using this model one can ask the following questions:

- How many rearrangement attempts/cell are compatible with the measured frequency of κ^+/κ^o among peripheral B cells?
- What values of *a* and *f* are compatible with the measured frequency of κ^+/κ^o among peripheral B cells?
- In order to accommodate the notion of massive secondary light chain rearrangements is it necessary to postulate that κ alleles do not rearrange independently, but rather that multiple rearrangement attempts preferentially occur on a single chromosome?

In order to evaluate these parameters we have plotted the cumulative % κ^+/κ^o among functional cells that are generated after each round of rearrangement attempts in terms of the entire range of values of *a* and *f* (Fig. 3). As can be seen from Fig. 3, only maximal allowed values of 3–5 rearrangement attempts/cell yield κ^+/κ^o percentages that are compatible with the published results. A maximal level of allowed rearrangements/cell of five is only compatible with high values of *a* and *f*, whereas a maximal level of allowed rearrangements/cell of three is only compatible with low levels of *f*, i.e., a high frequency of light chain non-functionality. A maximal level of allowed rearrangements/cell of four is compatible with all possible

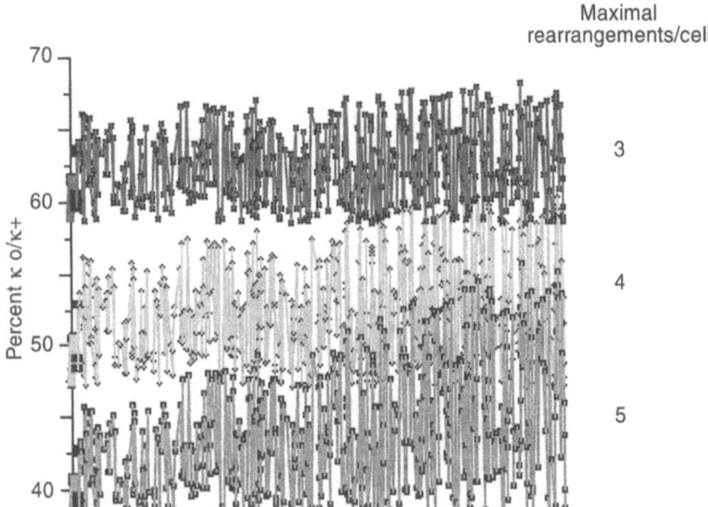

Fig. 3. Extensive editing because of autoreactivity would have only a minor effect on the frequency of κ^+/κ^o cells. Three to five rounds of rearrangement attempts are compatible with published percentages of κ^+/κ^o splenic B cells. Data points were generated with random values of *f* and *a* using the JMP statistical analysis program (SAS Institute, N.C.). Lines connecting points were included to distinguish different values of maximal allowed rearrangements

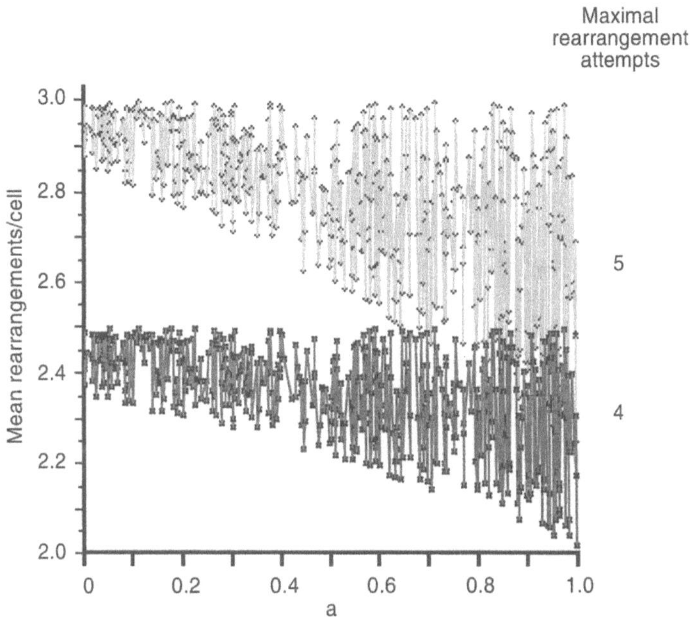

Fig. 4. A narrow range of mean rearrangement attempts/cell is compatible with known κ^+/κ^o: κ^+/κ^- ratios

f and *a* values. A maximum of four V_κ–J_κ rearrangements/cell corresponds in turn to the narrow range of 2.0–2.5 mean rearrangements/cell (Fig. 4).

3 Discussion

The present model suggests that massive secondary light chain rearrangement induced by either autoantigen-mediated receptor editing or light chain non-functionality is compatible with published percentages of κ^+/κ^o B cells. Assuming that up to four rearrangement attempts are allowed before RS/λ recombination is substantially initiated, then all values of *a* and *f* are consistent with random rearrangements on both chromosomes, and do not require preferential secondary rearrangements on previously rearranged chromosomes in order to explain the κ^+/κ^o data. More or fewer than a mean of 2.0–2.5 V_κ–J_κ rearrangements/cell severely constrains the possible values of *a* and *f* or would suggest a non-random κ allele usage.

It is clear that the assumption that κ rearrangements are distributed randomly upon both alleles is an oversimplification because it is possible that once a VJ rearrangement occurs at one locus, a second rearrangement at the same locus may be more likely, e.g., because of V region promoter-enhanced recombination accessibility. On the other hand, a second rearrangement at the same locus may be less likely because fewer Js and Vs are available on the rearranged locus than on an unrearranged locus. Little direct experimental data exists that would support either possibility. However, with eight functional J_κs/cell, mean values of 2–3 κ rearrangements/cell would not be significantly influenced by "using up" J_κs. In hemizygous κ-deficient mice the rate of κ-B cell production is only lowered by approximately 10% relative to κ-sufficient controls (Chen et al. 1993; Zou et al. 1993), consistent with a normal maximum number of κ rearrangements/cells of approximately four prior to initiation of RS (and λ) rearrangements.

The concept of light chain non-functionality in the present context predicts four categories of such light chain genes: (1) pseudogenes that bear stop codons, (2) pseudogenes that bear structural mutations precluding pairing with any heavy chain, (3) potentially functional genes whose protein products are structurally incompatible with some, but not all heavy chains, and (4) functional genes whose products are able to form allowable HL pairs, but whose mRNA expression level is abnormally low. This latter category may well be under-represented among rearrangements because the transcriptional activation of V_κ genes appears to be closely correlated with rearrangement efficiency (Stiernholm and Berinstein 1995), therefore it seems unlikely that the V_κ genes that can rearrange lack functional promoters. The frequency of light chains of category 3 above is unknown, but a high frequency would negate or reduce the advantages of H/L combinatorial diversity.

In summary, the available evidence is compatible with the idea that receptor editing is a major mechanism during B cell development. More detailed studies are needed to determine the extent to which tolerance and light chain non-functionality contribute to secondary light chain rearrangements.

Acknowledgements. This work was supported by grants from the National Institutes of Health (RO1AI33608, K04AI01161) and the Arthritis Foundation. The author thanks Drs. Prak, Litwin and Weigert for commenting on the manuscript. This paper was also presented in a volume for the International Symposium on Immune Tolerance organized by the Fondation Mérieux.

References

Chen J, Trounstine M, Kurahara C, Young F, Kuo C-C, Xu Y, Loring JF, Alt FW, Huszar D (1993) B cell development in mice that lack one or both immunoglobulin λ light chain genes. EMBO J 12: 821–830

Chen C, Radic MZ, Erikson J, Camper SA, Litwin S, Hardy RR, Weigert M (1994) Deletion and editing of B cells that express antibodies to DNA. J Immunol 152:1970–1982

Clarke S, McCray S (1991) A shared κ reciprocal fragment and a high frequency of secondary $J_\kappa 5$ rearrangements among influenza hemagglutinin specific B cell hybridomas. J Immunol 146:343–349

Coleclough C, Perry RP, Karjaleinen K, Weigert M (1981) Aberrant rearrangements contribute significantly to the allelic exclusion of immunoglobulin gene expression. Nature 290:372–378

Doglio L, Kim JY, Bozek G, Storb U (1994) Expression of λ and κ genes can occur in all B cells and is initiated around the same pre-B cell developmental stage. Dev Immunol 4:13–26

Durdik J, Moore MW, Selsing E (1984) Novel kappa light-chain gene rearrangements in mouse lambda light chain-producing B lymphocytes. Nature 307:749–752

Feddersen RM, Van Ness BG (1985) Double recombination of a single immunoglobulin κ-chain allele: implications for the mechanism of rearrangement. Proc Natl Acad Sci USA 82:4793–4797

Gay D, Saunders T, Camper S, Weigert M (1993) Receptor editing: an approach by autoreactive B cells to escape self tolerance. J Exp Med 177:999–1008

Gollahon KA, Hagman J, Brinster RL, Storb U (1988) Ig λ-producing B cells do not show feedback inhibition of gene rearrangement. J Immunol 141:2771–2780

Harada K, Yamagishi H (1991) Lack of feedback inhibition of $V_κ$ gene rearrangement by productively rearranged alleles. J Exp Med 173:409–415

Hardy RR, Dangl JL, Hayakawa K, Jager G, Herzenberg LA, Herzenberg LA (1986) Frequent λ light chain gene rearrangement and expression in a Ly-1 B cell lymphoma with a productive κ chain allele. Proc Natl Acad Sci USA 83:1438–1442

Huber C, Klobeck HG, Zachau HG (1992) Ongoing $V_κ$–$J_κ$ recombination after formation of a productive $V_κ$–$J_κ$ coding joint. Eur J Immunol 22:1561–1565

Kwan SP, Max EE, Seidman J, Leder P, Scharff MD (1981) Two kappa immunoglobulin genes are expressed in the myeloma S107. Cell 26:57–66

Langman RE, Cohn M (1995) The proportion of B-cell subsets expressing κ and λ light chains changes following antigenic selection. Immunol Today 16:141–144

Levy S, Campbell MJ, Levy R (1989) Functional immunoglobulin light chain genes are replaced by ongoing rearrangements of germline $V_κ$ genes to downstream $J_κ$ segments in a murine B cell line. J Exp Med 170:1–13

Lewis S, Rosenberg N, Alt F, Baltimore D (1982) Continuing kappa-gene rearrangement in a cell line transformed by Abelson murine leukemia virus. Cell 30:807–816

Ma A, P. Fisher P, Dildrop R, Oltz E, Rathbun G, Achacoso P, Stall A, Alt FW (1992) Surface IgM mediated regulation of RAG gene expression in Em-N-myc B cell lines. EMBO J 11:2727–2734

Moore MW, Durdik J, Persiani DM, Selsing E (1985) Deletions of κ chain constant region genes in mouse λ chain-producing B cells involve intrachromosomal DNA recombinations similar to V-J joining. Proc Natl Acad Sci USA 82:6211–6215

Nemazee D (1996) Antigen receptor "capacity" and the sensitivity of self-tolerance. Immunol Today 17:25–29

Prak EL, Weigert M (1994) Light chain editing in κ-deficient animals: a potential mechanism of B cell tolerance. J Exp Med 180:1805–1815

Prak EL, Troustine M, Huszar D, Weigert M (1995) Light chain replacement: a new model for antibody gene rearrangement. J Exp Med 182:541–548

Radic MZ, Erikson J, Litwin S, Weigert M (1993) B lymphocytes may escape tolerance by revising their antigen receptors. J Exp Med 177:1165–1173

Russell DM, Dembic Z, Morahan G, Miller JFAP, Bÿrki K, Nemazee D (1991) Peripheral deletion of self-reactive B-cells. Nature 354:308–311

Shapiro MA, Weigert M (1987) How immunoglobulin $V_κ$ genes rearrange. J Immunol 139:3834–3839

Stiernholm BJN, Berinstein NL (1995) A mutated promoter of a human Ig Vλ gene segment is associated with reduced germ-line transcription and a low frequency of rearrangement. J Immunol 154:1748–1761

Tiegs SL, Russell DM, Nemazee D (1993) Receptor editing in self-reactive bone marrow B-cells. J Exp Med 177:1009–1020

Tonegawa S (1983) Somatic generation of antibody diversity. Nature 302:575–581

Van Ness BG, Coleclough C, Perry RP, Weigert M (1982) DNA between variable and joining gene segments of immunoglobulin κ light chain is frequently retained in cells that rearrange the κ locus. Proc Natl Acad Sci USA 79:262–266

Verkoczy LK, Stiernholm BJN, Berinstein NL (1995) Up-regulation of recombination activating gene expression by signal transduction through the surface Ig receptor. J Exp Med 154:5136–5143

Zou Y-R, Takeda S, Rajewsky K (1993) Gene targeting in the $Ig_κ$ locus: efficient generation of λ chain-expressing B cells, independent of gene rearrangements in $Ig_κ$. EMBO J 12:811–820

Clone: A Monte-Carlo Computer Simulation of B Cell Clonal Expansion, Somatic Mutation, and Antigen-Driven Selection

M.J. SHLOMCHIK[1], P. WATTS[2], M.G. WEIGERT[3], and S. LITWIN[2]

1 Introduction

During clonal expansion of antigen (Ag)-stimulated B lymphocytes, the immunoglobulin (Ig) variable region (V) genes undergo point mutation at a high rate (McKEAN et al. 1984; CLARKE et al. 1985). Somatic diversity results from the

[1]Yale University School of Medicine, Department of Laboratory Medicine, Box 208035, 333 Cedar St. New Haven, CT 06520-8035, USA. mark.shlomchik@yale.edu
[2]Fox Chase Cancer Center, Department of Biostatistics, 7701 Burholme Avenue, Philadelphia, PA 19111, USA
[3]Princeton University, Department of Molecular Biology, Schultz Laboratories, Princeton, NJ 08544, USA

sequential accumulation of such mutations (CLARKE et al. 1985; McKEAN et al. 1984; CUMANO and RAJEWSKY 1986; JACOB et al. 1991b). These mutations, and selection of them by Ag, have important consequences for the kinetics, quality, and size of the resulting immune response. Hence, somatic mutation is a dynamic system and in this sense is distinct from mechanisms that shape the preimmune repertoire such as combinatorial joining and gene conversion.

The nature of somatic mutation has been revealed by sequencing V regions of Abs isolated at various times after immunization (CLARKE et al. 1985; McKEAN et al. 1984; CUMANO and RAJEWSKY 1986; JACOB et al. 1991b; GRIFFITHS et al. 1984; APEL and BEREK 1990; ZIEGNER et al. 1994). From these sequence data estimates have emerged of rate (McKEAN et al. 1984; CLARKE et al. 1985), inference of both positive and negative selection of mutations (SHLOMCHIK et al. 1990; CLARKE et al. 1985), and limited insight into the mutational mechanism (JACOB et al. 1993). Nonetheless, analysis of sequence data has generally not considered all the available information on the biological process. As a result, a wealth of sequence information exists which, we believe, could reveal much more about the dynamics of B cell clonal expansion and selection if analyzed more thoroughly.

For such analysis, we have developed a computer model of clonal expansion. By tuning the parameters of our model to fit available data on the frequencies and patterns of mutation, we can precisely describe the selective forces acting on B cells, determine the upper and lower limits on mutation rates, and measure the variation and magnitude of the response to different Ags. Here we will use the model to determine the mutation rate, estimate the size of the combining site (i.e., the extent of sequence that can be positively selected), and demonstrate how selection during clonal expansion influences the frequencies and patterns of mutation. Most importantly, we will provide ranges for these values within which experimental data may be interpreted, with particular emphasis on the inherent variation in the stochastic process of clonal expansion and selection and how this affects our perception of the process.

2 The Nature of Somatic Mutation

2.1 Sequential Accumulation

Direct evidence for mutation came first from the observation that members of a clone of B cells had sequence differences (McKEAN et al. 1984). Moreover, mutations within (but rarely between) sets of clonally related B cells are extensively shared, and the sharedness is hierarchical. Accumulation of mutation is also found in comparisons of Abs early and late in the immune response (JACOB et al. 1993). That point mutations accumulate is seen from the frequencies of mutations in Abs at early stages of diversification (JACOB et al. 1991b, 1993). Also, numerous examples of clonally related Abs that differ by one base substitution have been observed (e.g., SHAN et al. 1994). [There are sites at which it appears that two-base changes happen (see Sect. 2.3) but these are unusual].

2.2 Mutation Rate

The average rate at which somatic mutations accumulate has been estimated to be 10^{-3}/bp per generation (gen) (MCKEAN et al. 1984; CLARKE et al. 1985). This estimate is based on the mutational content of cells isolated at a certain time after the onset of mutation and the number of cell divisions during that time span. The onset of mutation is delayed; in spleen, it appears that mutations do not appear before approximately 1 week (JACOB et al. 1993). Also, rates sharply decrease later (SHLOMCHIK et al. 1987b; SIEKEVITZ et al. 1987; CLAFLIN et al. 1987). Our model incorporates the initial lag but in this version does not model the effect of mutation stopping.

2.3 Mutation Pattern

Patterns are interpreted in the context of two functional categories of residues in Ig V regions: framework regions (FRs) and complementarity determining regions (CDRs). FRs were originally defined as contiguous stretches of primary sequence which showed relatively little variability from Ab to Ab (WU and KABAT 1970; KABAT 1966, 1967, 1970). KABAT postulated that FRs function as the relatively constant framework upon which the more variable CDRs are supported. The CDR regions, due to their high variability, were thought to contact Ag, hence the term "complementarity determining". This hypothesis has been borne out by the study of X-ray derived crystal structures of Ab–Ag complexes (MACCALLUM et al. 1996). Prior to nucleotide sequencing (and the enumeration of both silent and replacement mutations) replacements were found to be concentrated in CDRs (WEIGERT et al. 1970; CESARI and WEIGERT 1973). This is particularly marked for Abs at early stages of diversification. At this stage all replacements were in CDRs (WEIGERT et al. 1970; CESARI and WEIGERT 1973). The analysis of V region protein sequences from plasmacytomas expressing members of the $V_{\kappa}21$ family provided a sampling of sequences at later stages of diversification (VALBUENA et al. 1978). Although R mutations were found at a higher frequency in CDRs, mutations in FRs, presumably neutral replacements, were also found. Such patterns can also be deduced from the genealogies of mutations in clones. Mutations at early stages of clonal expansion (i.e., those shared by all clone members) are usually Rs in CDR (e.g., CLARKE et al. 1985).

A more interesting pattern is the R/S ratio as measured across the V region. This can be significantly high in CDRs – as expected for positive selection (R/S ratios are correspondingly low in FRs as expected for negative selection) (SHLOMCHIK et al. 1987a, b, 1990; CLARKE et al. 1985). This pattern is seen not just for somatic mutation but also for meiotic mutation (CHANG and CASALI 1994,

1995). Evidence of the latter comes from the codon bias in germline V genes. These often lead to CDR nucleotide sequences that favor R mutations, for example the SER codon AGy is used in CDR, and the alternative TCx/y is used in FR. It should be noted that this bias can influence the expected R/S ratio in CDR (Chang and Casali 1994). This is important since the significance of the R/S ratio is usually based on a comparison to the base line R/S ratio, i.e., that expected in the absence of positive or negative selection. This expected value will vary according to codon usage and must be calculated to assess properly the significance of a somatically derived R/S ratio (as described by Casali and collaborators; Ikematsu et al. 1993a, b; Chang and Casali 1994).

The R/S ratio can vary considerably. Although at early stages of clonal expansion the R/S ratio is more likely to be high, at later stages it may or may not be significantly high. A likely determining factor is the potential of a CDR for improved binding to a particular Ag. However, the reasons for R/S ratio variation are complex and one goal of modeling is to understand the basis of variation.

Somatic mutations are found near the V region promoter and extend to the downstream J regions; hence, with regard to the entire genome, mutations are clustered (Weber et al. 1994). However, within the V(D)J region there is no evidence of clustering. That is, silent mutations do not appear to center on small subregions of the V, such as in CDRs. There is, however, a slight tendency to introduce two adjacent mutations (D.S. Smith et al. 1996), which is thought to relate to the error-prone mechanism of synthesis or repair inherent in the mutagenesis. Such pairs, though more frequent than predicted at random, are nonetheless rare even in highly mutated sequences.

2.4 Hotspots

Another type of inhomogeneity of the mutation process is the existence of hypermutable sites or hotspots. Various hotspots have been identified (Berek and Milstein 1987; Insel and Varade 1994). Those which do not involve amino acid substitutions (and thus are silent or S mutations) must result from features of the mutagenesis mechanism. These have also been revealed in "passenger Ig V transgenes" that are not expressed (Betz et al. 1993). There are certain sequence motifs which appear to be mutated more frequently than others (Insel and Varade 1994; Jacob et al. 1993). However, importantly for our analysis, these hotspots are not concentrated in one part of the V gene or another. Thus, although both dinucleotide changes and hotspots exist, they do not invalidate the general concept that mutations are distributed in a random manner over the average V region.

2.5 Selection

Even at hypermutation rates, the frequency of any given higher affinity mutant will be low. Intense positive selection must be involved to explain their expansion. As discussed in Sect. 2.3, R/S ratios that are significantly higher than expected by chance are common and also demand intense selection. How such selection is manifested and which phenotypes are selected are major unknowns. As discussed in Sect. 5, this model begins to define these parameters of the immune response.

3 Somatic Mutation and Burst Size

Clone sizes have been estimated by extrapolation from the frequencies of related B cells captured in a fusion. These extrapolations depend on highly variable values such as fusion efficiencies and are imprecise (SHLOMCHIK et al. 1987a; SHAN et al. 1994). The identification of a site in which mutation occurs during the primary response now gives a more precise estimate. Using a combined morphologic and polymerase chain reaction (PCR) sequencing approach, JACOB and KELSOE identified the germinal center (GC) as the major site of mutation during the primary immune response (JACOB et al. 1991b). The microdissection/PCR technique used in these studies also allowed a better determination of mutation kinetics and led to rough estimates of the number of mutating B cells at various points in the immune response (JACOB et al. 1991b, 1993).

These and other studies (APEL and BEREK 1990; ZIEGNER et al. 1994) have led to the following general picture of the primary immune response and GC reaction. Upon initial Ag exposure, B cells become partially activated and migrate to the outer periarteriolar lymphoid sheath (PALS) zone. There, they encounter Ag-specific T cells, to which they presumably present Ag; this encounter leads to both a T and B cell proliferative response. This response peaks at 3–4 days and has largely disappeared by day 7. Some of the B cells proliferating in the PALS give rise to plasma cells, which migrate to the red pulp of the spleen and possibly to the bone marrow (LIU et al. 1991; DILOSA et al. 1991). These provide the early Ab response. At some point during this PALS response, an unknown number of B and T cells begin to migrate to the GC, where they begin rapid proliferation. GCs first become noticeable at around days 5–7 of an immune response and continue to expand for 1–2 weeks thereafter. Mutated B cells in the PALS have never been observed, and mutants from early GC reactions are rare. Thus it is assumed that somatic mutation initiates sometime early in the GC reaction, possibly at its start. Ag in the form of immune complexes accumulates on follicular dendritic cells (FDC) (NOSSAL et al. 1965; TEW et al. 1980). These FDCs are peculiar to the GC, and have been considered a source of ligand for centrocytes that might promote B cell selection (TEW et al. 1980; MACLENNAN 1994). By this model, Ab already bound to the Ag would compete with surface Ig on B cells; this competition would permit an effective signal only to higher affinity B cells.

4 Questions That can be Addressed by Modeling

Thus GC studies have refined our understanding of the dynamics and extent of mutation. In turn, they raise new questions. What is the schedule and rate of mutation? Does mutation occur only in certain cells/sites in the GC, and does selection occur only in certain cells/sites? What is the impact of positive and negative selection on the final repertoire? How much selection and what size selection target is required to account for high R/S ratios and affinity maturation (i.e., how many mutations are selectable)? What is the nature of competition within a GC and between GCs?

Computer simulations provide us with a group of plausible solutions that in turn can be tested experimentally. Simulations also help us to quantify factors inherent in the data that are not easily accessible by simple statistics. This is accomplished because Monte Carlo simulations are essentially descriptions of the data or an average of the data. In this regard, they resemble descriptive statistics. However, they can embody properties of the complex and dynamic system that static statistics do not. A good example is selection and its effects. It is difficult to conceptualize selection (much less deduce its strength or extent) by examination of the raw data; but selection becomes a quantifiable term in a computer simulation. In this regard, a simulation also functions as a framework or syntax for thinking about and understanding the process being simulated. We hope to demonstrate some of these principles here.

5 Model Basics and Assumptions

Our model begins with a single B cell dividing under the influence of positive and negative selection. This cell might represent one that was stimulated in the outer PALS a few days before and now has arrived in the GC to begin the process of mutation, proliferation, and selection. We assume, as discussed in Sects. 2.3 and 2.4, that somatic mutation is a random process. The rate (a variable called *mut* in the model) can be set at any value. The onset of hypermutation can be instantaneous, or the mutation rate can increase gradually over any set number of divisions. In the runs presented here, mutation is set to begin at the onset, as in the case of a B cell proliferating in a GC. (The phenomenon of mutation stopping is modeled similarly, although it is not included here.)

A fundamental assumption of this model is that cell fates depend on the type of mutation. There are four types of mutation: neutral, lethal, disadvantageous (but not lethal), and advantageous. Neutral mutations include all silent (S) mutations; the frequency of S mutations in Ab V genes is usually approximately 0.25 and is modeled as a constant. R mutations that are neither advantageous nor lethal are also considered neutral; these are catalogued but have no effect on the cell. The

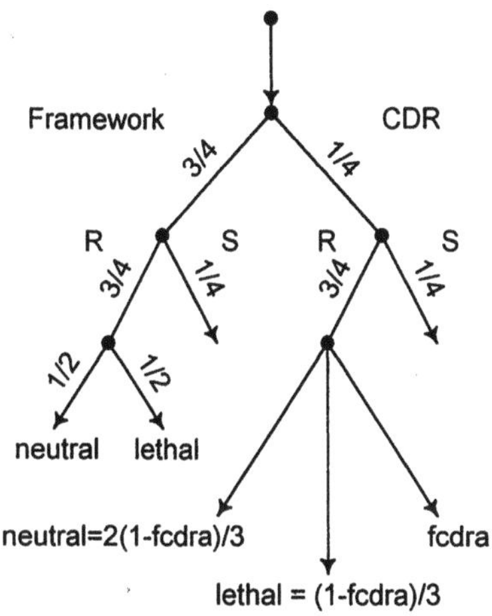

Fig. 1. Logic diagram of cell fates and mutation in the model. Cells are subjected to a Poisson-distributed number, n, of mutations at each division. The logic of this figure is exercised n times for such a cell and the outcomes summed. The path to a neutral framework replacement mutation has probability $3/4 \times 3/4 \times 1/2 = 9/32$, which also holds for a lethal framework mutation. The chance of an advantageous mutation is $1/4 \times 3/4 \times fCDRa = 3fCDRa/16$, and that of a complementarity-determining region (*CDR*) lethal mutation is $1/4 \times (1-fCDRa)/16$. *fCDRa* is a parameter set at runtime. If n is at least one, a uniformly distributed random number, μ, is selected. For example, if μ is less than 9/32 then a framework neutral replacement mutation is assigned the cell's daughter. If μ exceeds 9/32 but is less than 18/32, a framework lethal mutation is assigned. Other mutation types are treated similarly. New mutations are added to those inherited from the mother cell. If any lethal mutations were added, the daughter cell is deleted from the run

frequency of various types of mutations and how they affect cell fate in the model is given in Fig. 1.

Other R mutations will be lethal, such as mutations that destroy the folding of the Ab molecule. In Ab V genes, the frequency of such lethal mutations can be estimated by analyzing the patterns of variability at each amino acid position among large collections of diverse Ab sequences. Most of these lethal mutations occur in the FRs of Abs. Nevertheless, the FR is moderately variable, although the amino acid substitutions seen are mainly conservative. Based on the observed variability in FR we calculated that about half of R mutations in FR would be lethal (SHLOMCHIK et al. 1990). This in turn predicted that the R/S ratio in FRs should be 1.5, a value that has now been reported for FR mutations in a wide variety of Ab genes (INSEL and VARADE 1994). Thus, we use 0.5 for the fraction of lethal R mutations in FRs (fFRl) and 0.5 for the fraction of FR neutral R mutations.

Certain R mutations will be advantageous (CDRa-type mutations). The number of CDRas will vary depending on the functional size of the combining site and is thus modeled as the variable (*fCDRa*), which is the fraction of the CDR R mutations that are advantageous. The advantage or disadvantage a cell has compared to its siblings depends on the types of mutations it has acquired. We represent selective advantage as a decreased time interval between divisions. (Other conceptions are possible: for example, selective advantage could also be modeled if division rates were constant but cells with selective advantage had a decreased chance of cell death.)

The key variable for determining selective advantage is the division time of each individual cell (T_i), which is determined from several other variables and parameters as follows: The maximum interval between divisions for a cell which is still specific for Ag is defined as T_{max}; this presumes that the minimum affinity for Ag required to trigger a B cell leads to division every T_{max} hours. Cells with no advantageous mutations (including the starting cell) divide every T_{max} hours. T_{min} is the shortest division time and applies to a cell that is "optimized" for binding to a particular Ag. T_{min} is a parameter which represents the minimum amount of time in which any B cell can complete a cell cycle. T_{min} is usually set at 6–8 h, in accordance with experimental data. The difference between T_{max} and T_{min} (which is referred to as Δ) is the window of time within which positive selection can operate. As cells accumulate advantageous CDR replacement mutations (RCDRa), their division times (defined for cell "i" as T_i) will approach T_{min}. The extent to which a mutation decreases division time (T_i) depends on the weight given each mutation. We have modeled this relationship as follows:

$$T_i = T_{min} + \Delta/[\mathrm{Ag}(t)s^{a_i}],$$

where a_i is the number of RCDRa for the ith cell. This is tallied in the computer for each cell. s, stands for selection factor. This is a scalar value set at the beginning of the run that embodies the selective weight that each successive *RCDR*a mutation confers upon the cell. The value of s is raised to the a power and this quantity is divided into Δ, thus reducing the difference between a cell's minimum and maximum division times proportionally. When s is small, for example 1.1, then selection is minimal, since 1.1^a will be a small number and thus not shorten division time significantly. When s is large, a single RCDRa mutation will reduce the cell's dividing time to nearly its minimum (i.e., T_{min}).

The function describing T_i has several notable features. Most prominent is that it assigns the greatest selective value to the early mutations (Fig. 2). Clearly, not every initial advantageous mutation will have this result in vivo. However, we do believe that cells will sense differences in affinity of their sIg to a greater degree when at a relatively lower affinity; whereas cells already at a high affinity are not likely to respond much better if that affinity is improved. Moreover, as advantageous mutations accumulate in vivo, the likelihood of additional advantageous mutations decreases. We do not explicitly reduce the chance of advantageous mutations in our model (i.e., the parameter *fCDRa*) since the nature of the T_i function embodies this. The most obvious alternative to the current T_i function is to randomly apportion the decrements in Δ (which is the window of division time optimization) to the individual RCDRa. This will generally result in a somewhat less optimal enrichment of RCDR mutations; however, we suspect this effect will be small since cells which happen to get a large apportionment at the start will rapidly dominate during exponential growth. The issue of different T_i functions, including the use of models which control the probability to die rather than divide, remains to be investigated.

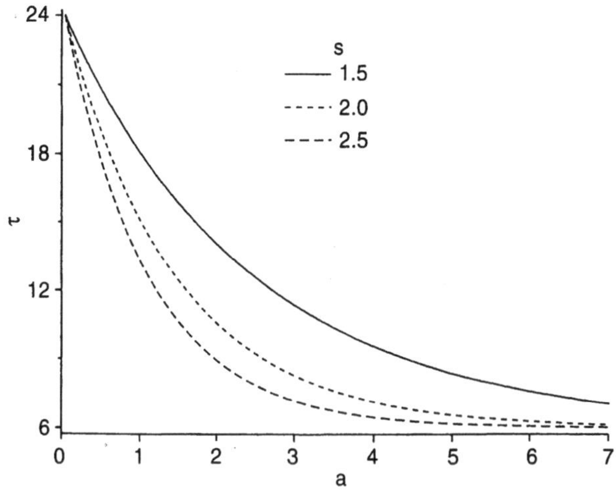

Fig. 2. How division time *tau* (T_i in the text) is affected by increasing numbers of advantageous (a) (x-axis value) mutations. Cell generation time, τ, is determined by three runtime parameters T_{min}, T_{max}, and s (see text), as well as by the number of advantageous mutations, a, inherited by a cell at its time of birth, and the level of Ag, Ag(t), at this time. Ag level is determined by t and two additional fixed parameters (see Fig. 3). Curves are shown for three different settings of s. Ag levels are held constant in these curves. (See Fig. 3 for effect of Ag concentration)

Ag depletion is controlled by the function Ag(t):

$$\mathrm{Ag}(t) = 1 - \frac{\mathrm{K} - \Phi[(t - t_{1/2})/\sigma]}{2\mathrm{K} - 1}$$

for $0 < t < 2t_{1/2}$ and $\mathrm{Ag}(t) = 0$ for $t > 2t_{1/2}$, where $\mathrm{K} = \Phi[-t_{1/2}/\sigma]$ and Φ is the normal distribution function.

The clearance of Ag most certainly controls the duration and vigor of the immune response. This function allows one to use a variety of clearance (decay) curves. There are two parameters which control the shape of the curve: $t_{1/2}$, the half-time of Ag decay, and σ, the standard deviation of the Ag decay (in essence, the slope of the curve during the accelerated decay phase). Figure 3 illustrates a family of decay curves with different $t_{1/2}$ and σ values. Here we have chosen a sigmoidal curve so that decay at first will be slow (due to lack of Ab) and then will rapidly accelerate as Ab is formed during the middle part of the immune response. Late in the response, the rate of clearance will again slow due to limitations in [Ag]. We have not investigated other monotonic functions of Ag decay (e.g., linear or exponential) but suspect from our results that the model is not dramatically sensitive to the shape of the curve; it is at least somewhat sensitive to the $t_{1/2}$ of the [Ag], a parameter which we have investigated (see Sect. 6.5.2).

Our Ag depletion function is not designed to correlate clearance with Ab affinity; it differs in this sense from other models (CELADA and SEIDEN 1992; KEPLER and PERELSON 1993, 1995).We think such "feedback loops" are inappropriate because most Ab in the primary response is made at around day 4 from

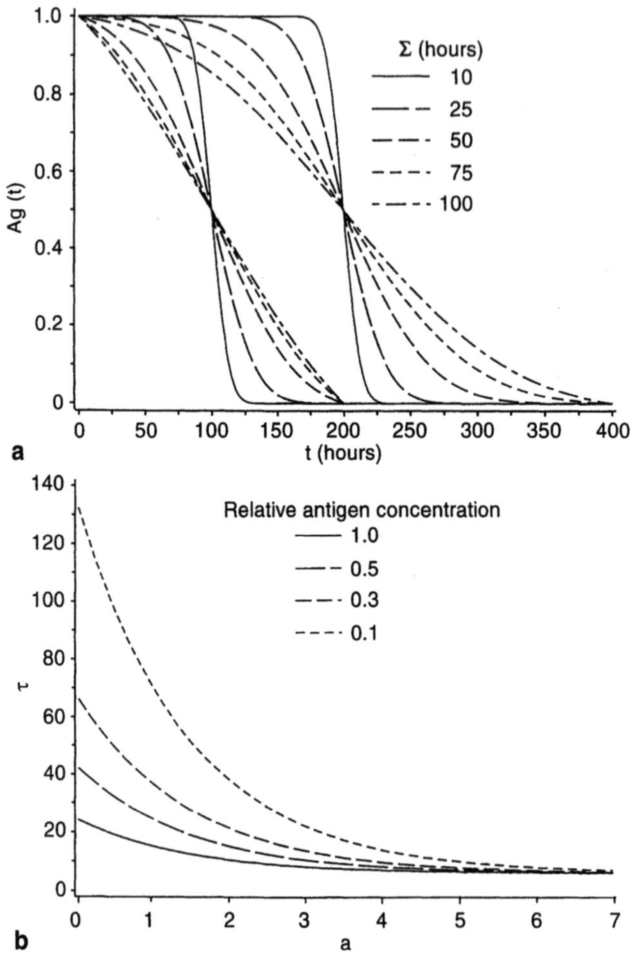

Fig. 3a, b. How relative Ag levels affect division time and how Ag is controlled in the model. Ag is depleted using a two-parameter sigmoidal function (see text). $t_{1/2}$ is Ag half-life and σ determines the speed of depletion near the half-life point. **a** Two families with half lives of 100 h and 200 h (*left*, *right*) with $s = 10, 25, 50, 75, 100$ h with 10 h shown as *solid line* $\mathrm{Ag}(t) = 0$ for $t > 2t_{1/2}$. **b** The effect of Ag depletion on generation time with $s = 2$, $t_{1/2} = 100$ h and $\sigma = 50$ h

descendants of B cells that have proliferated in the outer PALS (Han et al. 1995; Smith et al. 1996). Thus, we envision that the early primary immune response (e.g., days 4–14) will be comprised of a clonally diverse set of germline-encoded, low affinity Abs. Since Abs can freely move about, we suspect that every cell in every GC will be exposed to the same competitive Ab. The affinity/avidity of this germline Ab should be uniform and independent of the success or affinity of the B cells in the GC (which are not thought to secrete much if any Ab at this time). Of course, the situation is much different in a secondary immune response, and we would adopt new assumptions and models to account for this.

6 Results

6.1 Parameter Values That Simulate the Outcomes of Primary GC Reactions

The main features of the experimental data that constrain our model are: numbers of mutations, patterns of mutations (i.e., R/S values), and numbers of cells ("burst size"). All of these are available from published data and will be discussed in Sects. 6.2 and following. A priori, three kinds of outcomes of the parameter search were possible. At one extreme, we might find that no combination of model parameter values can give us reasonable results. This would require us to reevaluate the assumptions inherent in our approach. An intermediate result would be that some parameter values give credible results, but that the values required are "unphysiological". An example would be finding that a mutation rate of $< 10^{-4}$/bp/gen is optimal, even though the actual rate is approximately 10^{-3}/bp/gen. Another example would be an *fCDRa* approaching 1.0, meaning that every R mutation in CDR would have to be advantageous. The outcome hoped for would be that sets of parameter values lead to model outcomes that are consistent with the experimental data constraints and with physiological parameter values. Admittedly, the distinction between the latter outcome and the intermediate outcome is somewhat vague, since the physiologic values may not truly be known. It is here where we submit that modeling can direct us to specific experiments or more careful examination of existing data to clarify these issues. In this sense, modeling becomes a data interpretation tool. Similarly, the values of certain parameters such as *s* will give us insight into forces that shape the immune response, but for which we do not have a syntax or meaning. Again, we expect such model discoveries to lead to experimental insights and to provide conclusions as to the way the GC reaction works.

To summarize our results, it was possible to find sets of model parameters that both satisfied known experimental data (the constraints) and seemed physiological. We will describe the process we used to find such values, discuss insights gained from these values and future work that could be done in this modeling system.

6.2 Parameter Search Methods

We began by selecting a fairly broad set of parameter values which we considered to be physiologically plausible. These are summarized in Table 1. We focused on day 10 of the immune response, which approximates to GC day 12. This assumes that the GC is seeded on day 2 with a few cells and becomes noticeable by days 4–5 after several divisions of these seeding cells. Most of the detailed information we have about mutation and GC size come from analyses of day-10–16 GCs (Jacob et al. 1991a, b, 1993; Jacob and Kelsoe 1992; Ziegner et al. 1994; Apel and Berek 1990). Obviously, other time points in the model remain to be studied. The initial broad set of parameter values was executed in the model 20 times each. (Each set of

Table 1. Parameter values examined in three consecutive sets of runs

Run set	Total number of runs	*mut*	T_{max}	*fCDRa*	*t*	*s*	$t_{1/2}$	σ (fraction of $t_{1/2}$)
1	192	0.5, 1	24, 48[a]	0.1, 0.25, 0.5	504, 1004	1, 1.5, 2, 2.5	0.5[b] t	0.1, 0.2, 0.5, 1
2	216	0.5, 1, 1.2, 1.5	24	0.1, 0.25, 0.5	252, 348	1.5, 2, 2.5	0.5[b] t	0.2, 0.5, 1
3	192	0.5, 0.8	22, 24	0.25, 0.5	252, 288	2, 2.5	0.5[b] t, 0.67[b] t	0.2, 0.5, 1

For each run set, every combination of the values listed in each cell was run for ten iterations (set 1) or 20 iterations (sets 2 and 3). This combinatoric yielded the indicated total number of runs.
[a]When $T_{max} = 48$, t was set to 1004 and when $T_{max} = 24$, t was set to 504.
[b]$t_{1/2}$ was always set to a fraction of the total run time (t), as indicated.

parameter values is termed a "run", while each trial of the same parameter values is referred to as an "iteration" of the run.) An initial screening was performed by ranking all the runs in terms of the mean R/S ratio in the CDR.

Many runs generated an average burst size (over all the iterations) that was too high (> 2000) or too low (< 100). These cutoffs are based on estimates of the number of cells in a day-12 GC (about 1300, calculated from Jacob et al. 1991a). It is unlikely that parameter sets with average bursts fewer than 100 cells (i.e., comprising less than 8% of GC cells) would contribute greatly to the experimental data at hand. In other words, low burst sizes are essentially invisible at our current experimental resolution since in most cases less than ten examples from a single GC or whole spleen have been studied. High burst sizes have not been further considered because these would tend to create peak GC sizes well before the observed peak at days 10–12. In addition, they would generally create GCs that contain too many cells altogether, though we cannot rule out other limiting factors that could constrain such clones once they reach maximum GC size (such as limiting T cell help or constraints imposed by the follicle size or number of FDCs).

A final filtering was performed based on the total number of mutations. Again, we rely on data of Jacob and Kelsoe which demonstrate an average of six mutations per cell at day 10 (Jacob et al. 1993). The standard deviation of this value is unknown; arbitrarily we considered runs that generated between three and nine average total mutations per cell to be plausible.

The parameter values of the remaining small number of runs were used to generate a larger number of iterations in order to gain confidence in the mean values. A prominent feature of the results was wide variability in burst size. We suspected that different burst sizes would have different biologic consequences, and thus we analyzed the results by burst size as follows: 0–100, 101–500, 501–2000, and > 2000 cells. This exercise was revealing in that there was a wide range of burst sizes from different iterations using the same parameter values. This is illustrated for two sample runs in Table 2. Such burst size variation was seen with each realistic parameter value set run and is, we believe, an inherent feature of B cell clonal

Table 2. Variation in burst size between iterations as indicated by two separate runs

Run	Burst size category	R/S ratio in CDR	Burst per iteration	Iterations in burst size category (%)	Total mutations per cell
14	All	5.3	185	100	7.9
	0–< 100	2.6	46	86	2.6
	100–< 500	5.3	211	10	5.0
	500–2000	6.9	939	3	7.3
	2001 +	5.4	7107	1	10.8
24	All	5.6	2890	100	10.8
	0–< 100	2.3	61	25	3.0
	100–< 500	4.4	216	34	5.0
	500–2000	5.5	1056	16	7.5
	2001 +	5.6	10854	24	11.2

In run 14, $mut = 0.5$, $T_{max} = 22$, $fCDRa = 0.25$, $t = 252$, $s = 2.5$, $t_{1/2} = 126$, $\sigma = 63$. Run 24 was similar except: $fCDRa = 0.5$, $t_{1/2} = 168$, $\sigma = 168$. R/S ratios, as well as total mutations per cell, vary greatly between iterations that fall into each burst size category, even though the run parameter values were identical for each iteration. Note that between runs 14 and 24, within a burst size category, both R/S and total mutations are comparable. However, the percentages of iterations falling into each burst size category are quite different, with run 24 generating many more large iterations and hence a much greater average burst per iteration when considering all the iterations. Even though run 24 generated a quarter of all iterations with excessive numbers of cells, a quarter of its iterations still generated less than 100 cells.

expansion itself (see Sect. 6.3). Given this, we considered all other parameters in the context of burst size. We again discarded runs that generated a substantial fraction (> 10%) of iterations with > 2000 cells and selected runs that generated high R/S values with appropriate numbers of total mutations. Among the selected runs, we only considered (for purposes of this R/S analysis) iterations that generated between 100 cells and 2000 cells. Even more (750) iterations of this select group of runs (Table 3) were generated to increase the precision of our conclusions; these data are presented and discussed in Sects. 6.3 and following. In addition, select runs were performed that were related in some way to our ideal runs for comparison purposes and to assess the influence of changing a specific parameter on the results.

6.3 Burst Size Variation

Perhaps the most striking and unexpected feature of our simulation data is the dramatic variation in burst size from iteration to iteration of the same input parameters. This variation was seen with all of the parameter sets examined as illustrated in Fig. 4. Several conclusions can be drawn from this result. First, among runs which generate > 2000 cells with considerable frequency (e.g., > 20% of the time), most of the iterations still generate fewer than 100 cells. Even the most lifelike parameter values, i.e., those that generate the requisite number of total mutations and R/S ratios, still generate iterations of less than 100 cells. The fundamentally great variance in the burst size is independent of chosen model parameters and is an inherent feature of clonal expansion with random mutagenesis along with positive and negative selection. Therefore, we conclude that the in vivo immune response in a GC is likely seeded by more precursors than had originally been thought, but most of these never yield detectable numbers of progeny. Thus, while a GC is oligoclonal when well developed (see KROESE et al. 1987; JACOB et al. 1993), its initial seeding is likely to be multiclonal. A hint of this point is found in the data

Table 3. Best runs

Family of Runs	Run	Burst size category	$t_{1/2}$	σ	R/S ratio in CDR	Iterations in burst size category (%)	Total mutations per cell
A							
$T_{max} = 22$	7	100–500	126	25	6.1	17	5.0
$fCDRa = 0.5$		501–2000			7.8	2	7.2
$s = 2$	8	100–500	126	63	5.2	19	5
		501–2000			7.6	3	7.9
	9	100–500	126	126	6	21	5
		501–2000			7.4	4	7.6
	10	100–500	168	34	4.7	48	4.6
		501–2000			6.7	14	7.5
	12	100–500	168	168	5.3	45	4.6
		501–2000			6.7	13	7.5
B							
$T_{max} = 22$	14	100–500	126	63	5.3	10	5
$fCDRa = 0.25$		501–2000			6.9	3	7.3
$s = 2.5$							
C							
$T_{max} = 22$	19	100–500	126	25	5.2	22	5.3
$fCDRa = 0.5$		501–2000			5.6	4	7.9
$s = 2.5$	20	100–500	126	63	6.2	21	5.6
		501–2000			6.9	8	8.5
	21	100–500	126	126	5.7	19	5.4
		501–2000			6.5	8	8.4
	22	100–500	168	34	4.9	42	4.8
		501–2000			7.2	17	8.0
	23	100–500	168	84	4.9	37	5.4
		501–2000			5.8	15	7.1
	24	100–500	168	168	4.4	34	5
		501–2000			5.5	16	7.5
D							
$T_{max} = 24$	55	100–500	126	25	7.1	8	5.1
$fCDRa = 0.5$		501–2000			11	1	7.6
$s = 2$	56	100–500	126	63	7.9	10	5.5
		501–2000			10	3	8.3
	57	100–500	126	126	7	12	5.7
		501–2000			11	2	7.9
	58	100–500	168	34	4.7	44	4.4
		501–2000			6.3	9	7.3
	59	100–500	168	84	5	30	5.1
		501–2000			6.4	9	7.6
	60	100–500	168	168	5.9	26	5.2
		501–2000			6.6	10	7.7
E							
$T_{max} = 24$	64	100–500	210	42	3.6	57	4.2
$fCDRa = 0.25$		501–2000			4.8	13	6.7
$s = 2.5$	65	100–500	210	105	4.2	40	4.4
		501–2000			6	9	6.9
	66	100–500	168	168	4.1	21	4.9
		501–2000			5.1	7	7

Table 3. (*Contd.*)

F							
$T_{max=24}$	67	100–500	126	25	6.2	16	5.6
$fCDRa = 0.5$		501–2000			5.8	4	7.9
$s = 2.5$	68	100–500	126	63	6	12	5.4
		501–2000			6.9	5	8.0
	69	100–500	126	126	7.1	14	6.1
		501–2000			5.5	6	8.2

For brevity, only the data from burst size categories between 100–2000 cells are given. All runs shown used $mut = 0.5$, and $t = 252$. All runs in a family also shared the indicated values of T_{max}, $fCDRa$, and s. Each run within a family differed in $t_{1/2}$ and/or σ, as indicated. Note that differences in σ had little impact, which makes the number of independent solutions smaller than would appear. Note that the families themselves also have highly related parameter values, compared to the range that was examined (see Table 1).

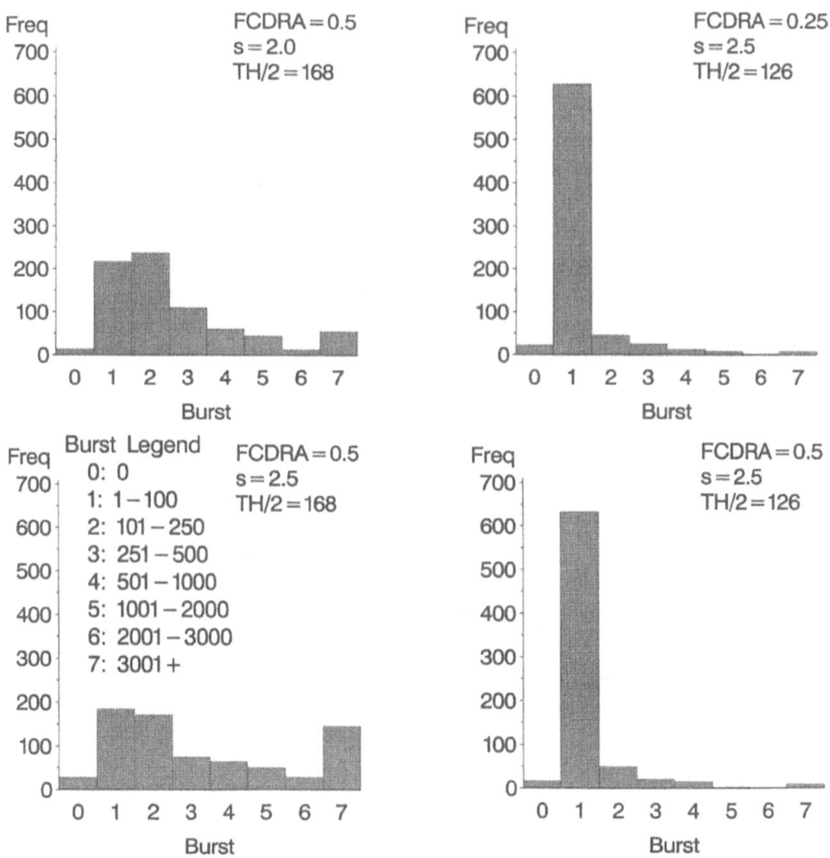

Fig. 4. Distribution of burst size (i.e., the number of total live cells when the iteration ended at day 10) for several parameter combinations. Seven burst size categories were defined as shown in the legend on the figure. Iterations were assigned to a category depending on the number cells they generated. All four runs used $mut = 0.5$ and $t = 252$ h. Values for $fCDRa$, s and $t_{1/2}$ ($TH/2$) are given above each histogram. A total of 750 iterations were run for each set of parameter values. *Freq*, the number of iterations with burst in a particular bin

of Jacob and Kelsoe, who noted much greater diversity (six clones on average) in nascent GCs than in late GCs (Jacob et al. 1993). (Higher diversity was also suggested in chronic human GCs by the results of a single cell analysis (Kuppers et al. 1993) although in this case the diversity seen could result from chronic re-seeding of an ongoing GC.) "Clonal failure" on this scale is a heretofore under-appreciated source of inefficiency in the immune system. We had appreciated that negative selection would lead to slower clonal growth (Shlomchik et al. 1990). It had not been anticipated that a large proportion of starting clones would never generate significant numbers of progeny; this characteristic of clonal selection in the GC may be crucial for maintaining a reasonable number of responding B cells.

We have termed those clones which generate few progeny clonal failures, but they actually do generate some progeny. These could be sources for memory B cells and serve to populate secondary responses. It is tempting to speculate that such cells are the source of clonal shifts that occur between primary and secondary responses. In clonal shifts, the clonotypes that predominate in secondary responses are rarely, if ever found in the primary responses. The many small clones generated in a GC that have not been well appreciated up to now are candidates for the precursors of dominant clones in the secondary, whose origins so far have not been identified. However, we doubt this is generally the case. This is because the clonal failures also have low numbers of mutations and low R/S. That is to say, the reason they fail is lack of early CDRa mutations. This is the opposite of the situation in dominant secondary clones which share early mutations. Furthermore, we doubt that clones that have not achieved higher affinity due to CDRa mutations, as is the case with clonal failures, will be able to compete with clones that have achieved higher affinity as well as higher precursor frequency.

6.4 Mutation Rate

We have found upper and lower limits for the rate of somatic mutation. The upper limit is mainly constrained by negative selection on mutations in FR regions that affect conserved residues and thus would inactivate V regions (these are FRl mutations; we have set the parameter that controls this value, fFRl, to 0.5, as discussed in Sect. 5). At too high a value of *mut*, the frequency of inactivating mutations will become high enough to prevent clonal expansion. On average this will occur when *mut* approaches 1.6–1.8 (depending on other parameter values). Even as *mut* approaches one, the average burst size for many parameter value sets was much less than 100 cells. However, some parameter value sets with a *mut* of one have an average burst per iteration of 1000 or more. Analysis of these data by iteration, however, revealed that this high average number was because of a few rare iterations which generated many cells. It might be argued that these rare cases are biologically significant, as in the case of clonal failures and successes discussed in Sect. 6.3. However, analysis of the total number of mutations in these rare clonal successes at $mut = 1$ shows that those parameter value sets that generated on average > 300 cells per iteration had total numbers of mutations of 15–25 per cell.

This is much greater than observed values and thus allows us to rule out the biological significance of these runs.

What is the optimal rate? In our analysis, it is about 0.5. This was the *mut* value for all of the most optimal runs identified as described in Sect 6.2. Rates much lower, such as 0.25, would neither generate enough mutations nor enough cells (due to lack of positive selection). A *mut* rate of 0.5 corresponds to about 10^{-3}/bp/gen, also the best available rate estimate in vivo. These estimates were made under a number of assumptions that rendered them subject to a several-fold error. Indeed, others have suggested rates as low as 10^{-4}/bp/gen (BEREK and MILSTEIN 1987). But variations of even two-fold or less (e.g., *mut* = 1 vs *mut* = 0.5) give quite different results in our model. This suggests that our current picture of mutation and selection is quite sensitive to a particular mutation rate that does not tolerate much variation. Our model is thus useful both in supporting the previous rate estimate of 10^{-3}/bp/gen (MCKEAN et al. 1984; CLARKE et al. 1985) and in demonstrating that the range of the actual rate will be close to this estimate.

6.5 Parameter Values That Generate High R/S: What They Mean and How They Function

We have just demonstrated how the mutation rate is critically constrained. What about other variables in our model? We have tried to understand the importance, function, and probable values of these variables using two basic approaches: first, by comparing parameter value sets among the optimal ones identified as above; and second, by titrating single parameter values in the context of the other values of some of these optimal sets. We will consider each of the parameters in turn.

6.5.1 Parameters T_{max} and σ

In the ranges we checked, we did not find much dependence on these variables. In the case of T_{max}, this is likely because we did not vary this parameter value much (checking 22 h and 24 h only). T_{max} embodies the inherent fitness of the germline V genes. As such, it most certainly will vary among different clones and different immune responses. However, in our case, we wished to simplify the situation by considering how other variables would affect the response given the same inherent fitness of the starting B cell. T_{max} is also related to the affinity cutoff that determines whether a B cell is functionally "specific" for a given Ag or else does not sense that Ag. In the latter case, T_{max} is infinite. We suspect that a cutoff exists in that once a cell can sense the Ag physiologically, it begins to divide at a certain minimum rate; nonetheless, this rate is unknown. We have guessed that for an average V region, this rate is once a day. By varying this a few hours, we have shown that this time does not have a major impact on the other variable values or results. However, we have not yet explored much higher or lower values for T_{max}.

σ was also not found to affect the results in a significant or predictable way. This parameter controls the shape of the sigmoidal curve of Ag depletion. Little is actually known about the change of effective Ag concentrations with time during the immune response. The fact that output was not particularly sensitive to the shape of this curve suggests that the assumption of sigmoidal Ag depletion is not a critical one. Even if Ag depletion follows a somewhat different (though monotonic) depletion function we expect that results will not be much affected.

6.5.2 The $t_{1/2}$ Parameter

This parameter determines the half-life of Ag depletion. Depending on other parameter values, we found that there was broad variability in the acceptable values for $t_{1/2}$. On the other hand, we did not obtain good results when the $t_{1/2}$ was either too short or too long; best results were obtained when $t_{1/2}$ was set somewhere near the midpoint of the GC reaction. If $t_{1/2}$ is too small, then burst sizes are too small and the number of mutations are also too small (see Table 4); the response literally dies out. This is not surprising since we presume that, regardless of the unknown actual kinetics of effective Ag concentration, the GC reaction remains Ag-driven and thus the concentration must remain at a biologically significant level till at least the peak of the reaction. A $t_{1/2}$ that is too long gave the opposite result of too many iterations with large burst sizes and with a large number of mutations. These

Table 4. Effect of antigen $t_{1/2}$ on burst size

Run	Burst size category (hundreds of cells)	Ag[a] $t_{1/2}$	R/S ratio in CDR	Burst per iteration	Iterations in burst size category (%)	Total mutations per cell
569	All	72	4.3	11	100	1.8
	0–< 100		4.1	11	100	1.8
	100–< 500		11.3	103	0	4.2
56	All	126	7.2	177	100	10.1
	0–< 100		3.6	35	86	2.6
	100–< 500		7.9	204	10	5.5
	500–2000		10.2	792	3	8.3
	2001 +		7.1	7705	1	13.6
59	All	168	7.2	742	100	10.3
	0–< 100		3.0	53	55	2.9
	100–< 500		5.0	214	30	5.1
	500–2000		6.4	1036	9	7.6
	2001 +		7.5	9764	6	11.8
568	All	210	6.1	1249	100	9.5
	0–< 100		2.6	61	27	3.1
	100–< 500		4.5	221	48	4.7
	500–2000		6.1	956	15	7.1
	2001 +		6.2	9772	10	10.5

[a] All runs differed only by the $t_{1/2}$ value, with σ set to half of the $t_{1/2}$.
All runs shared the following parameter values: $mut = 0.5$, $T_{max} = 24$, $fCDRa = 0.5$, $t = 252$, $s = 2$. Note that with very low $t_{1/2}$, few runs even achieve > 100 cells, whereas with longer $t_{1/2}$, a significant percentage of runs generate > 2000 cells, with the average generating nearly 10 000. There was, on the other hand, little effect on the total numbers of mutations within each category. Note also, that for maximal R/S, a moderate value of $t_{1/2}$ was best ($t_{1/2} = 126$).

iterations were not as optimal with regard to R/S ratios either, suggesting the need for reduced Ag concentrations to drive affinity selection. It should be pointed out the there is no one correct $t_{1/2}$, but that it will depend on the Ag, dose, and probably the genetic capacity of the responding immune system. Our results also predict properties of $t_{1/2}$ that may correlate with observed immune responses. For example, prolific responses with large numbers of mutation without great evidence of Ag-selection (i.e., without high R/S) may be indicative of persistent Ag. This pattern is actually seen in surgically removed human tonsils, which are likely chronically infected with persistent pathogens (Liu et al. 1996; Pascual et al. 1994).

6.5.3 The *fCDRa* Parameter

This parameter, which gives the fraction of R mutations in CDR that lead to selective advantage, is a function of the Ab/Ag pair. An Ab can have any *fCDRa*, depending on its structure and that of its Ag. As we will show, only if *fCDRa* is substantial will high R/S be detectable. For this reason high R/S will not be found in all types of responses. Some V regions may be nearly perfect to begin with, perhaps due to evolutionary selection by a ubiquitous pathogen. The anti-phosphorylcholine response in inbred mice seems to fall into this category. Other Ag/CDR interactions may be limited to just a few contacts. Since high R/S is readily observable in a variety of immune responses this suggests that *fCDRa* must be large in many settings. We have used the model to estimate the minimum *fCDRa* that could regularly lead to detectably high R/S ratios in CDRs. *fCDRa* of 0.25 leads to high R/S ratios but only when combined with certain other parameter values, such as high *s* (see Sect. 6.5.4). A wider range of other parameter values is compatible with high R/S when *fCDRa* is 0.5. One test of the model's validity is whether such high values of *fCDRa* are plausible. At least for protein Ags, we believe they are, based on the number of contacts that an Ab makes with Ag as determined by crystal structures. A recent summary by MacCallum et al. of 26 Ag/Ab pairs shows that, for large Ags, nearly half of all CDR residues make contact with Ag (MacCallum et al. 1996). As expected, the fraction is lower for smaller Ags, though still considerable. If only half of all available mutations lead to an improvement, this would mean an *fCDRa* of about 0.25.

6.5.4 The *s* Parameter

The parameter *s* is probably the most difficult one to deal with. It embodies assumptions and simplifications concerning how affinity-increasing mutations affect cell division times. Nonetheless, the values that optimize *s* can give us considerable insight into the magnitude of effect that the average CDRa mutation must confer in order to lead to high R/S ratios that often characterize Ag-selected Abs. We find that *s* needs to be 2–2.5 to lead to such R/S values. Indeed, when *fCDRa* is 0.25, *s* must be 2.5 or higher to lead to a substantial fraction of iterations with high R/S.

Table 5. Effect of *s* on R/S ratio in CDR, total mutations and burst size

Run	Burst size category (hundreds of cells)	*s*[a]	R/S ratio in CDR	Burst per iteration	Iterations in burst size category (%)	Total mutations per cell
567	All	1.25	3.0	30	100	2.3
	0–< 1		3.0	30	100	2.3
566	All	1.5	3.9	36	99	2.7
	0–< 1		3.4	34	97	2.6
	1–< 5		9.0	129	2	5.0
56	All	2	7.2	177	100	10.1
	0–< 1		3.6	35	86	2.6
	1–< 5		7.9	204	10	5.5
	5–20		10.2	792	3	8.3
	20+		7.1	7705	1	13.6
565	All	2.5	8.3	815	100	11.4
	0–< 1		4.1	35	76	2.7
	1–< 5		6.0	192	12	5.4
	5–20		6.9	893	5	8.0
	20+		8.4	9948	7	12.1
564	All	3	6.6	1875	100	12.1
	0–< 1		3.3	35	65	2.8
	1–< 5		5.9	219	13	6.5
	5–20		6.8	1015	7	8.7
	20+		6.6	11044	16	12.4

[a] All runs differed only by the *s* value.
All runs shared the following parameter values: $mut = 0.5$, $T_{max} = 24$, $fCDRa = 0.5$, $t = 252$, $t_{1/2} = 126$, $\sigma = 63$.

Certain interesting properties of *s* emerge from examining how output varies when *s* is varied while holding other parameter values steady, as shown in Table 5. As *s* is lowered from optimal values around 2, the number of iterations that yield at least 100 cells drops. For example, in the case of run 56 and its variants, at $s = 2$, 14% of iterations yield > 100 cells, whereas at $s = 1.5$, only 2% do. Interestingly, the R/S of the iterations with substantial numbers of cells actually increases as *s* decreases. Again, using the run 56 example, R/S is 6.8, 6.9, 10.2, 9.0 as *s* is 3, 2.5, 2.0, and 1.5, respectively. When *s* is raised to 3, the most striking result is the fraction of iterations that yield very large numbers of (> 2000) cells, which is 16% for run 56, compared to 1.4% when $s = 2$. The bell-shaped dependence of R/S on *s* is seen by the fact that these > 2000 cell iterations have an R/S of only 6.6 and the 500–2000 cell iteration category has R/S of 6.8. Compare this to $s = 2$ in which case the 500–2000 cell category has R/S of 10.2. The net result of this dependency is that, when all cells are considered, R/S usually peaks at mid-range values of *s* (see Table 4). There are two reasons why moderate values of *s* are optimal: each mutation must have a substantial effect on division time to permit its enrichment during clonal expansion; when *s* is too low this does not occur. On the other hand, there must be incremental steps of selection to allow high R/S ratios, which in turn result from selection of several independent RCDRa mutations. If *s* is too high this does not occur.

7 Conclusions

Our clonal proliferation model is valuable in a number of respects. First, it has helped to refine our understanding of mutation rate and the fraction of CDR regions that are susceptible to improvement in an average anti-protein Ab response. These points were implicit in accumulated data but required a simulation approach to be understood. Similarly, we have gained insight into how Ag clearance might affect ongoing immune responses. These results make predictions about Ag clearance, although it will be difficult experimentally to determine the kinetics of "available" (i.e., unbound) antigenic epitopes during the course of an immune response. In any case, we show that the shape of the clearance curve is probably not as important as the $t_{1/2}$ of the Ag. Finally, we have investigated one function that relates the number of advantageous mutations to cell fate. Although any construct such as this is artificial, we have reached the fundamental conclusion that the selective advantage must be split into several (not too few and not too many) finite parts in order to achieve high R/S ratios. This is essentially the meaning of the finding that s must be neither too low nor too high. It also seems likely that early mutations will have to be given more weight than subsequent ones, on average. This is inherent in our function; we are in a position to test whether other arrangements can also readily yield successful simulations with high R/S ratios.

A major feature of our model is the high degree of variability from iteration to iteration. As we have already argued, it is very likely that this is a feature of clonal expansion of somatically mutating B cells in vivo. One of the most variable characteristics is cell burst size. Thus, depending on the chance occurrence of advantageous mutations or not, the same rules of selection (i.e., parameter values) can generate anywhere from zero to thousands of cells in 10 days. In every parameter value set we studied, iterations of < 100 cells were frequent. We could have tuned the parameters to eliminate this, but it would have resulted in nearly all of the iterations generating many more cells than we know exists in the day 10 GC. Thus, we feel clonal failures are a common and inherent part of clonal proliferation. This in turn leads to the idea that GC may be seeded by many more than previously thought (Jacob et al. 1993; Kroese et al. 1987), with only a minority of these being successful. The fate of the multiple small clones in contributing to memory responses is uncertain, though we suggest that ordinarily they are not major contributors. On the other hand, there are inevitably iterations that generate large numbers of cells. It is not clear how or whether these extraordinary clones are dealt with in vivo. Perhaps they create unusually large GCs; little information is available as to how large a GC can become. Alternatively, perhaps the follicular structure, FDCs, or the availability of T cell help become limiting, which results in control of the expansion of these clones.

A second implication of variability in R/S regards the interpretation of experimental R/S data. It is important to note that all of our simulations included selection that was programmed into them. However, not all of the iterations resulted in high R/S values that would be considered as indicative of selection

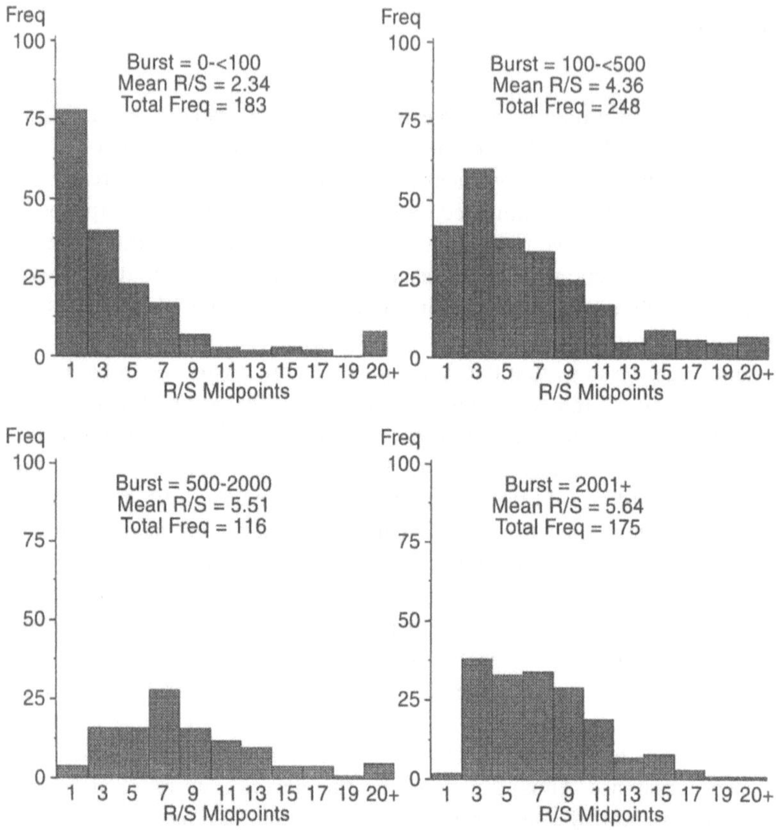

Fig. 5. Distribution of the CDR R/S ratio in run 10 (see Table 3 for parameter values). Of 750 iterations 736 produced at least one cell. Mean R/S within each of the 736 productive iterations was determined as follows: the sum of R mutations was divided by the sum of S mutations, where the summation was over all cells in the iteration's burst. This mean R/S was used to assign each iteration to an R/S group. The iterations were further divided into four groups according to burst size. The R/S distributions are shown individually for each burst size. Burst size groups are indicated above each of the four histograms. Bin midpoints rather than bin limits are used in the burst size group plots. *Freq*, the number of iterations with CDR R/S in a particular bin

(Fig. 5). On the other hand, when control runs were performed that did not include selection, high R/S ratios were almost never observed. What this means is that when high R/S is seen in experimental data, then it is quite reasonable to conclude that selection is the cause. However, when it is not seen, this in no way rules out that selection was nonetheless taking place. R/S and related measures, as indicators of selection, are thus shown to be highly specific but very insensitive. This should cause reevaluation of mutational analyses that purport to show an absence of selective processes; it is unlikely that reasonable power to detect selection was present.

Implicit in our results is that simple stepwise selection of cells with constant rates of mutation and expansion can yield simulations that are remarkably accurate in terms of cell burst, R/S ratios and total mutation. Thus, a priori, it is unnecessary to invoke a model of cyclic mutation and division. This issue will be dealt with in more detail elsewhere (SHLOMCHIK et al., in preparation).

8 Future Work

The model we have constructed and initially described here can be expanded and modified in several useful ways. One dimension of experimental data that we have not taken into account is "tree shape", or the genealogical relationships of clonal siblings. Different conditions would lead to different tree shapes, and tree shape, we believe, is an important indicator of the forces that shape clonal expansion. As yet, no suitable analytical tools exist to compare and quantify tree structure in a way useful to the analysis of GC B cell selection. Therefore, we plan to develop these and to include tree shape and its analysis as an additional constraint in the model. The T_i function that relates cell division time to the number of advantageous mutations has to be explored in greater depth. In particular, it would be interesting to model probability of death as opposed to probability to divide. Also, we can randomly assign steps of selective advantage instead of always making initial mutations provide the most relative advantage. Other issues yet to be explored are mutation stopping, which does seem to occur in certain circumstances (SHLOMCHIK et al. 1987; SIEKEVITZ et al. 1987; CLAFLIN et al. 1987) and is thought to have a major impact on clonal development and secondary immune responses. Overall, we are encouraged that our model is a useful tool both for data analysis and for generating insight into and predictions about somatic mutation and selection in B cell clonal expansion.

References

Apel M, Berek C (1990) Somatic mutations in antibodies expressed by germinal centre B cells early after primary immunization. Int Immunol 2:813–819

Berek C, Milstein C (1987) Mutation drift and repertoire shift in the maturation of the immune response. Immunol Rev 96:23–41

Betz AG, Rada C, Pannell R, Milstein C, Neuberger MS (1993) Passenger transgenes reveal intrinsic specificity of the antibody hypermutation mechanism: clustering, polarity, and specific hot spots. Proc Natl Acad Sci USA 90:2385–2388

Celada F, Seiden PE (1992) A computer model of cellular interactions in the immune system (review). Immunol Today 13:56–62

Cesari IM, Weigert M (1973) Mouse λ-chain sequences. Proc Natl Acad Sci USA 70:2112–2116

Chang B, Casali P (1994) The CDR1 sequences of a major proportion of human germline Ig V_H genes are inherently susceptible to amino acid replacement. Immunol Today 15:367–373

Chang B, Casali P (1995) A sequence analysis of human germline Ig V_H and V_L genes. The CDR1s of a major proportion of V_H, but not V_L, genes display a high inherent susceptibility to amino acid replacement. Ann NY Acad Sci 764:170–179

Claflin JL, Berry J, Flaherty D, Dunnick W (1987) Somatic evolution of diversity among anti-phosphocholine antibodies induced with Proteus morganii. J Immunol 138:3060–3068
Clarke SH, Huppi K, Ruezinsky D, Staudt L, Gerhard W, Weigert M (1985) Inter- and intraclonal diversity in the antibody response to influenza hemagglutinin. J Exp Med 161:687–704
Cumano A, Rajewsky K (1986) Clonal recruitment and somatic mutation in the generation of immunological memory to the hapten NP. EMBO J 5:2459–2468
Dilosa R, Maeda K, Masuda A, Szakal AK, Tew JG (1991) Germinal center B cells and antibody production in the bone marrow. J Immunol 146:4071–4077
Griffiths GM, Berek C, Kaartinen M, Milstein C (1984) Somatic mutation and the maturation of immune response to 2-phenyl oxazolone. Nature 312:271–275
Han S, Kathcock K, Zheng B, Kepler TB, Hodes R, Kelsoe G (1995) Cellular interaction in germinal centers. J Immunol 155:556–567
Ikematsu H, Ichiyoshi Y, Schettino E, Nakamura M, Casali P (1993a) V_H and V_κ segment structure of anti-insulin IgG autoantibodies in patients with insulin-dependent diabetes mellitus. J Immunol 152:1430–1441
Ikematsu H, Kasaian MT, Schettino EW, Casali P (1993b) Structural analysis of the V_H-D-J_H segments of human polyreactive IgG mAb. Evidence for somatic selection. J Immunol 151:3604–3616
Insel RA, Varade WS (1994) Bias in somatic hypermutation of human V_H genes. Int Immunol 6:1437–1443
Jacob J, Kelsoe G (1992) In situ studies of the primary immune response to (4-hydroxy-3-nitrophenyl)acetyl. II. A common clonal origin for periarteriolar lymphoid sheath-associated foci and germinal centers. J Exp Med 176:679–687
Jacob J, Kassir R, Kelsoe G (1991a) In situ studies of the primary immune response to (4-hydroxy-3-nitrophenyl)acetyl. I. The architecture and dynamics of responding cell populations. J Exp Med 173:1165–1175
Jacob J, Kelsoe G, Rajewsky K, Weiss U (1991b) Intraclonal generation of antibody mutants in germinal centres. Nature 354:389–392
Jacob J, Przylepa J, Miller C, Kelsoe G (1993) In situ studies of the primary immune response to (4-hydroxy-3-nitrophenyl)acetyl. III. The kinetics of V region mutation and selection in germinal center B cells. J Exp Med 178:1293–1307
Kabat EA (1966) Structure and heterogeneity of antibodies. Acta Haematol 36:198–238
Kabat EA (1967) Comparison of invariant residues in the variable and constant regions of human κ, human λ, and mouse κ Bence-Jones proteins. Proc Natl Acad Sci USA 58:229–233
Kabat EA (1970) Heterogeneity and structure of antibody-combining sites. Ann NY Acad Sci 169:43–54
Kepler TB, Perelson AS (1993) Somatic hypermutation in B cells: an optimal control treatment. J Theor Biol 164:37–64
Kepler TB, Perelson AS (1995) Modeling and optimization of populations subject to time-dependent mutation. Proc Natl Acad Sci USA 92:8219–8223
Kroese FGM, Wubbena AS, Seijen H, Nieuwenhuis P (1987) Germinal centers develop oligoclonally. Eur J Immunol 17:1069–1072
Kuppers R, Zhao M, Rajewsky K, Hansmann ML (1993) Detection of clonal B cell populations in paraffin-embedded tissues by polymerase chain reaction. Am J Pathol 143:230
Liu YJ, Zhang J, Lane PJ, Chan EY, MacLennan IC (1991) Sites of specific B cell activation in primary and secondary responses to T cell-dependent and T cell-independent antigens. Eur J Immunol 21:2951–2962
Liu YJ, de Bouteiller O, Arpin C, Briere F, Galibert L, Ho S, Martinez-Valdez H, Banchereau J, Lebecque S (1996) Normal human IgD^+ IgM^- germinal center B cells can express up to 80 mutations in the variable region of their IgD transcripts. Immunity 4:603–613
MacCallum RM, Martin AC, Thornton JM (1996) Antibody-antigen interactions: contact analysis and binding site topography. J Mol Biol 262:732–745
MacLennan ICM (1994) From the dark zone to the light. Curr Biol 4:70–72
Madnel TE, Phipps RP, Abbott A, Tew JG (1980) The follicular dendritic cell: long-term antigen retention during immunity. Immunol Rev 53:29–59
McKean D, Huppi K, Bell M, Straudt L, Gerhard W, Weigert M (1984) Generation of antibody diversity in the immune response of BALB/c mice to influenza virus hemagglutinin. Proc Natl Acad Sci USA 81:3180–3184
Nossal GJV, Ada GL, Austin CM, Pye J (1965) Antigens in immunity VIII. localization of ^{125}I-labelled antigens in the secondary response. Immunology 9:349–357

Pascual V, Liu Y-J, Magalski A, de Bouteiller O, Banchereau J, Capra JD (1994) Analysis of somatic mutation in five B cell subsets of human tonsil. J Exp Med 180:329–339
Shan H, Shlomchik MJ, Marshak-Rothstein A, Pisetsky DS, Litwin S, Weigert MG (1994) The mechanism of autoantibody production in an autoimmune MRL/lpr mouse. J Immunol 153:5104–5120
Shlomchik MJ, Litwin S, Weigert M (1990) The influence of somatic mutation on clonal expansion. Prog Immunol Proc 7th Int Cong Immunol 7:415–423
Shlomchik MJ, Aucoin AH, Pisetsky DS, Weigert MG (1987a) Structure and function of anti-DNA antibodies derived from a single autoimmune mouse. Proc Natl Acad Sci USA 84:9150–9154
Shlomchik MJ, Marshak-Rothstein A, Wolfowicz CB, Rothstein TL, Weigert MG (1987b) The role of clonal selection and somatic mutation in autoimmunity. Nature 328:805–811
Siekevitz M, Kocks C, Rajewsky K, Dildrop R (1987) Analysis of somatic mutation and class switching in naive and memory B cells generating adoptive primary and secondary responses. Cell 48:757–770
Smith DS, Creadon G, Jena PK, Portanova JP, Kotzin BL, Wysocki LJ (1996) Di- and trinucleotide target preferences of somatic mutagenesis in normal and autoreactive B cells. J Immunol 156:2642–2652
Smith KGC, Hewitson TD, Nossal GJV, Tarlinton DM (1996) The phenotype and fate of the antibody-forming cells of the splenic foci. Eur J Immunol 26:444–448
Valbuena O, Marcu KB, Weigert M, Perry RP (1978) Multiplicity of germline genes specifying a group of related mouse κ chains with implications for the generation of immunoglobulin diversity. Nature 276:780–784
Weber JS, Berry J, Manser T, Claflin JL (1994) Mutations in Ig V(D)J genes are distributed asymmetrically and independently of the position of V(D)J. J Immunol 153:3594–3602
Weigert MG, Cesari IM, Yonkovich SJ, Cohn M (1970) Variability in the λ light chain sequences of mouse antibody. Nature 228:1045–1047
Wu TT, Kabat EA (1970) An analysis of the sequences of the variable regions of Bence Jones proteins and myeloma light chains and their implications for antibody complementarity. J Exp Med 132:211–250
Ziegner M, Steinhauser G, Berek C (1994) Development of antibody diversity in single germinal centers: selective expansion of high-affinity variants. Eur J Immunol 24:2393–2400

[illegible]

Somatic Mutation in Ectothermic Vertebrates: Musings on Selection and Origins

L. Du Pasquier[1], M. Wilson[2], A.S. Greenberg[3], and M.F. Flajnik[3]

1 Introduction

The antibody responses of ectothermic (cold blooded) vertebrates do not mature in the same fashion as responses analyzed in mice. This has been demonstrated by either a total lack of affinity maturation in some species or a much lower rise in affinity in others (reviewed in Du Pasquier 1993). When it was determined that all ectotherms studied possess large numbers of V(D)J genes, and that rearrangement processes to establish Ig repertoires are essentially the same as those in mouse and human, it was theorized that poor immune responses in ectotherms could be explained by a suboptimal utilization of somatic mutants (Du Pasquier 1982, 1993). Another interpretation was that somatic mutation in the immune system arose late in vertebrate evolution, after emergence of the rearrangement process that generates functional V genes (Matsunaga 1985). Recent studies in cartilaginous fish (horned shark and nurse shark) and an amphibian (*Xenopus*) have shown conclusively that in addition to all of the molecular building blocks of the adaptive

[1]Basel Institute for Immunology, Grenzacherstrasse 487, Postfach 4005, Basel, Switzerland
[2]Department of Microbiology, University of Mississippi Medical Center, Jackson, MS 39216, USA
[3]Department of Microbiology and Immunology, University of Miami, P.O. Box 016960 R-138, Miami, FL 33101, USA

immune system, somatic hypermutation in immune-related genes is present in all jawed vertebrates. In this chapter we review the evidence for mutation in ectotherms, debate its importance to the immune system of these creatures, and speculate on its origins.

2 Evidence for Somatic Mutation in Ectothermic Vertebrates

2.1 *Xenopus* Studies

Xenopus is the only ectothermic vertebrate where somatic mutations have been followed during the course of an antigen-specific immune response (Wilson et al. 1992b, 1995). The V heavy chain (H) gene family involved in the anti-dinitrophenol (DNP) response was known (V_H1; Schwager et al. 1989) from previous sequencing of anti-DNP antibodies (Brandt et al. 1980); thus, one could monitor V_H1 gene expression during an ongoing anti-DNP response of isogeneic animals. First, all V_H1 genes (32 in total) present in a single haplotype were sequenced to establish an archive (Wilson et al. 1992a). These *Xenopus* genes and all of the other frog germline Ig genes, like their mammalian homologues, contain sequence motifs (A/G G C/T A/T) reported to target the mutational machinery (Schwager et al. 1989; Wagner et al. 1995).

The cDNA library was prepared with splenic mRNA from animals immunized 4 weeks earlier to DNP-KLH, when they were at the peak of their modest affinity maturation. Of the 32 members of the V_H1 family only five were expressed in an amplified manner, indicating that the immunization was being monitored. Among those genes corresponding either to IgM or to IgY (the IgG equivalent in *Xenopus*; Mussmann et al. 1996) only a small number of mutations were detected (an average of 1.6 mutations per gene, range: 1–5). There was not a strong preference for mutations in the complementarity determining regions (CDRs) 1 and 2, and virtually none were detected in CDR3 (Wilson et al. 1992b). While the frequency of mutations was lower than that found in responses of antigen-specific mammalian B cells, quantitative differences are equalized when one considers mutation rates. Taking the generation time of a *Xenopus* lymphoid tumor cell line as the basis of our calculation (Du Pasquier and Robert 1992), the rate of somatic mutation appears to be quite similar to that found in mammals during an antigen-specific response. The estimated *Xenopus* rates $-2.5'10^{-4}$/bp per cell generation to $4.1'10^{-5}$/bp per cell generation (average $1.5'10^{-4}$/bp per cell generation) – are only four to seven times lower than the highest levels reported in hyperimmunized mice.

In conclusion, the rate of appearance of somatic mutations in *Xenopus* probably does not account for the relatively poor affinity maturation of the antibody response as compared to mammals. Rather, a fundamental difference either in the mechanism or in the selection process is more likely in this system (see below).

2.2 Shark IgM and L Chains

Sharks mount an IgM antigen-specific response but show no affinity maturation (CLEM and LESLIE 1971; MÄKELÄ and LITMAN 1980). There is only one V gene family in cartilaginous fish which has 100–200 members (KOKUBU et al. 1987, 1988). Such homogeneity in a large number of V genes obviously hindered mutation studies in sharks until LITMAN and colleagues were able to uncover a single unique reference V_H germline gene (HINDS-FREY et al. 1993). Mutations in this gene were present at a slightly higher frequency than those in *Xenopus*. Putative somatic mutations also were found in the expressed germline joined V_L genes of another cartilaginous fish, the skate (ANDERSON et al. 1994, 1995). In both the shark and skate studies it was possible that some of the identified mutants were actually transcripts of other related but unidentified V_H genes. With this caveat in mind, the frequency and "type" of mutations (see below) were nevertheless indicative of a somatic hypermutation process. A mutation rate could not be calculated and no correlation with an immune response was attempted.

Shark and skate Ig genes are found in a so called cluster organization; for H chain genes each "cluster" contains one V gene, one or two D segments, one J, and one C gene (HINDS and LITMAN 1986). All evidence suggests that rearrangement occurs only within and not between clusters. Thus, LITMAN has proposed that diversity generated through somatic mutation preceded diversity obtained by combinatorial association of gene segments (HINDS-FREY et al. 1993).

2.3 Shark NAR

New or nurse shark antigen receptor (NAR) is an Ig-related molecule whose genes undergo rearrangement, which has been found so far only in sharks and rays (GREENBERG et al. 1995; RUMFELT and FLAJNIK, unpublished data). NAR is a disulfide-linked dimeric molecule (probably a homodimer) that does not associate with light (L) chains; each NAR molecule contains one V and five C domains. The number of NAR germline genes, all belonging to a single family, is relatively small (four to six genes) which permitted an analysis of somatic mutation, but as in the shark IgM study random cDNAs were studied. Figure 1 shows the incidence of hypermutation in 31 cDNA sequences encoded by one of the germline genes or a closely related allele. The frequency of mutations is much higher than for *Xenopus* and horned shark Ig, in fact much higher than in most studies in mammals. The mutations cluster relatively well in the CDRs (leader: 18 amino acid residues, 14 mutations; FR1: 25 residues, 64 mutations; CDR1: eight residues, 142 mutations; FR2: ten residues, 77 mutations; CDR2: ten residues, 112 mutations; FR3: 30 residues, 137 mutations), although mutations are found throughout the entire V region. It is extremely difficult to establish a pattern for the mutations because they are often contiguous and in large numbers. It is likely that the diversification is generated primarily through mutation and not gene conversion since the same pattern of changes is not detected in different cDNAs. However, a few cDNAs

```
                             A               B    -CDR1-     C     -CDR2-       D         E             F
     ├    Leader      ┤ ├──────────┤     ├───────┤        ├─────┤            ├──────┤   ├────┤        ├────
germ MNIFLLSVLLALLPYVFT ARVDQTPRSVTKETGESLTINCVLRDASYALGSTCWYRKKSGSTNEESISKGGRYVETVNSGSKSFSLRINDLTVEDGGTYRC
                                 10        20        30        40        50        60        70        80
1    ------------------ -----------------------------------------------------------------------------------
2    ------------------ -----------------------------------------------------------------------------------
3    ------------------ -----------------------------------------------------------------------------------
4    ------------------ ----------------------------------------P------------------------------------------
5    ------------------ ---A------------------A----RFDG-E---F------II--T--------------QE------K------S-----
6    -----------------E- -----------------------------E----------L------R-----------------C-----------------
7    ------------------ R--------------------------RSV-----------------------------I----A---T-----I--A-----
8    ------------------ --------------------------S---------D-------KYVG-VM--------------------------------
9    ------------------ G------------A----------H--R-G--R---G-PVPDR-YM-L--L----SQ-----------T--------S-----
10   ------------------ --------------------------V-LVK-G-----------K-----I-------------------------A------
11   -----------FV---W- ---------------------------N-G--R------SP---K----------------A-D--A------A---------
12   ------------------ ----------------------G----N-S-A---------------------------------------------------
13   ------------------ --------------------------GRLS-ER-----------G--P--RS------------F--------R---E-----
14   ///////////------- --L------------------------TD-EIRP----T----D----VR--D--------A-Q-------------A-----
15   ------------------ F--------------------------D----T----------P---R--P-------R--ER----------A---------
16   ------------------ ------------TE-------------TD---R-----Q-AD------------F------------------V---------
17   ------------------ --------------------S--------L--------S-L------R-E-----------------------VI------H-
18   ------------------ V-L----T---------------------T----------A------N-----------R---RS-S----------------
19   ------------------ ---------------------------N-P------F----------N--L-----------------------------F--
20   ------------------ ---------------------------N-------------------T---------------Q-------------------
21   ////////////////// /////----ER------------R---NN---R-----IR----------I---------K--N----------A--------
22   /////////--------- ---------------------------T---TN------------L---------------------------E-------H-
23   --------R--------- ---------------------------K----N-----TLQ---DD---E----F-----------------V----------
24   ////////////////// ////-------------V-L-------YH--NYN------P-G-S--RLPR-------F--A-D------T---P--------
25   ------------------ -------T----L----------I---TN--VR------L--R------E------------------------T--------
26   ------------------ V----------------------V-EL--------------------------------------------------------
27   ------------------ -------A-------D-------------D--E-----T----SD-KR-K--R------S--------T-K--VS--S-----
28   ------------------ --------T--R---------------N----G--------------I--GS----------T--------E-V---------
29   -----------------A -----------------V---------T-DVAN-F---------I------------RT-R-------T-KA-I-----N---
30   --V--------------- --L----------A---------------S------L----------R-----------------------------------
31   ------------------ ------K---------------------FGP-R------P---L-D----V----F--T------------------------
```

contain short sequence stretches that may have been derived from other germline loci or alleles, suggesting that conversion could also be operative in NAR genes. It is not known whether NAR genes diversify to generate the preimmune repertoire, or rather mutate in the course of an immune response. Preliminary data would favor the latter possibility (DIAZ and FLAJNIK, unpublished data).

The highest number of mutations are found in CDR1 (Fig. 2), which has the canonical targeting sequences mentioned earlier (e.g., see residue 28 and 33 in CDR1 in Fig. 2 and residue 48 in CDR2 in Fig. 3).

CDR 1	26	27	28	29	30	31	32	33	
Residue	D	A	S	Y	A	L	G	S	
germline	GAT	GCG	AGC	TAT	GCA	TTG	GGC	AGC	
1	---	---	---	---	---	---	---	---	
2	---	---	---	---	---	---	---	---	
3	---	---	---	---	---	---	---	---	
4	---	---	--T	---	---	---	---	---	0/1
5	---	---	--G	-T-	-AC	GG-	---	GAG	8/1
6	---	---	---	---	-A-	---	---	---	1/0
7	---	---	--A	AG-	-T-	---	---	---	4/0
8	---	T--	--T	---	--T	---	--G	---	1/3
9	---	---	--G	---	-G-	---	---	C--	3/0
10	---	-T-	---	-TA	-TT	AA-	--G	G--	7/2
11	---	---	-A-	---	-G-	---	---	C--	3/0
12	---	---	-A-	---	T--	---	-CA	---	3/1
13	---	-G-	--G	-TA	T--	---	-AG	--A	7/1
14	---	---	-CT	G-C	---	GA-	AT-	C--	7/2
15	---	--C	GAT	---	---	---	---	-C-	3/2
16	---	---	-CT	G--	--T	C--	---	C--	3/3
17	---	---	--T	--C	CTG	---	---	---	2/3
18	---	---	--T	--C	A--	---	---	---	1/2
19	---	---	-A-	---	C--	---	---	---	2/0
20	---	---	-A-	---	---	---	---	--T	1/1
21	---	---	-AT	A--	---	---	---	C--	3/1
22	---	---	-C-	---	---	---	ACT	-AT	4/2
23	---	---	-AA	---	--T	---	---	-AT	3/2
24	---	---	TA-	C--	--T	C-C	AA-	TA-	6/4
25	---	---	-C-	A--	---	---	-T-	C--	4/0
26	--G	CT-	---	---	--C	C--	---	---	3/2
27	---	---	--T	---	-AT	---	---	GAA	4/2
28	---	---	-A-	---	---	---	---	G--	2/0
29	---	---	-CT	---	-AC	G--	-C-	-A-	5/2
30	---	---	--T	--C	T--	---	---	---	1/2
31	---	---	TCG	-TC	-GG	CCT	--G	C--	7/5
	1R	5R	22R	13R	16R	9R	10R	21R	
	0S	1S	13S	5S	11S	5S	6S	4S	

Fig. 2. Nucleotide sequences of NAR complementarity-determining region (CDR) 1 and cDNA clones reveal a high frequency of mutation and canonical targeting sites. The amino acid residues are the same as in Fig. 1. The number of replacement (R) and silent (S) substitutions are indicated for each codon and each CDR. Note that silent substitutions often occur in codons in which a replacement mutation also is present. Total residues, 8; R/S ratio, 96:46 = 2.1

◄

Fig. 1. Amino acid sequences of random NAR cDNA clones reveals a high frequency of mutations. The germline sequence of the NAR leader and V region is shown in the *top line*. b strands and complementarity-determining regions are noted above this sequence. The bold residues in the germline sequence are conserved in most Ig/TCR sequences. The sequences reported here are a subset of those reported in GREENBERG et al. (1995)

CDR 2	44	45	46	47	48	49	50	51	52	53	
Residue	T	N	E	E	S	I	S	K	G	G	
germline	ACA	AAC	GAG	GAG	AGC	ATA	TCG	AAA	GGT	GGA	
1	---	---	---	---	---	---	---	---	---	---	
2	---	---	---	---	---	---	---	---	---	---	
3	---	---	---	---	---	---	---	---	---	---	
4	---	---	---	---	---	---	---	---	---	---	
5	-T-	-TT	---	---	-CG	---	--A	---	---	---	4/2
6	---	---	---	---	--A	---	---	---	---	---	1/0
7	---	---	---	---	---	---	---	---	---	---	
8	---	--A	T-T	-TT	G--	---	GTT	-TG	---	--C	9/3
9	--T	T-T	AT-	---	TTG	---	---	TT-	---	---	8/2
10	---	--A	---	---	---	---	---	-T-	---	---	2/0
11	---	--G	---	---	--T	---	--C	---	---	---	1/2
12	---	---	---	---	---	---	---	---	---	---	
13	--T	GG-	---	---	CCT	---	---	-GG	A--	---	6/3
14	GAT	---	---	---	--T	G--	CG-	--G	--C	-AC	5/6
15	C-C	---	---	---	C--	---	---	CC-	---	---	4/1
16	---	---	---	---	---	--C	---	---	---	---	0/1
17	---	---	---	---	--G	--T	GA-	---	---	---	3/1
18	---	---	---	---	-A-	---	---	---	---	---	1/0
19	---	---	---	---	---	---	---	CT-	---	---	2/0
20	---	---	---	---	-C-	---	---	---	---	---	1/0
21	---	---	---	---	---	---	--C	-T-	---	---	1/1
22	---	---	CT-	---	---	---	---	---	---	---	2/0
23	--G	G--	--C	---	--T	---	GA-	---	--G	---	4/3
24	---	TCG	---	---	--G	C--	C-A	-G-	---	---	7/1
25	---	---	---	---	---	---	GA-	---	---	---	2/0
26	---	---	---	---	---	---	---	---	---	---	
27	-GC	G--	---	A--	--G	--C	AA-	---	--C	C--	7/3
28	---	---	---	---	-T-	---	---	GG-	TC-	---	5/0
29	---	-T-	---	---	--T	---	---	---	---	---	1/1
30	---	---	---	---	--G	---	---	---	---	---	1/0
31	CT-	---	--C	---	---	---	---	GT	--A	---	5/1
	7R	13R	8R	3R	17R	2R	13R	15R	3R	2R	
	6S	2S	0S	0S	5S	3S	5S	3S	4S	2S	

Fig. 3. Nucleotide sequences of NAR complementarity-determining region (CDR) 2. Note the "targeted" *AGC* (Ser-48). The low level of mutation at Gly-52 and Gly-53 codons indicates that these residues have been selected in NAR for structural reasons, probably to form a bend between the two b sheets. Total residues, 10; R/S ratio, 82:30 = 2.73

3 Affinity Maturation and Selection: What Role Does Mutation Play?

The occurrence of mutations in Ig of ectotherms at frequencies and rates similar or higher than those of mammals is paradoxical in relation to the previously described low antibody heterogeneity and poor affinity maturation (Fig. 4). To account for this paradox, it was suggested (Wilson et al. 1992b) that mutated B cells are "poorly selected" in lymphoid organs of *Xenopus* and other cold blooded vertebrates because germinal centers, well known for their role in orchestrating the generation and selection of mutants in mouse and human (MacLennan 1994), are apparently absent in those species (Zapata et al. 1995, 1996a, b). If true then *Xenopus*, shark, and skate mutants would bear the hallmarks of this poor selection. One strong indication that this might be the case was that in spite of some clustering of the mutations in the CDRs, the ratio of replacment/silent (R/S ratio)

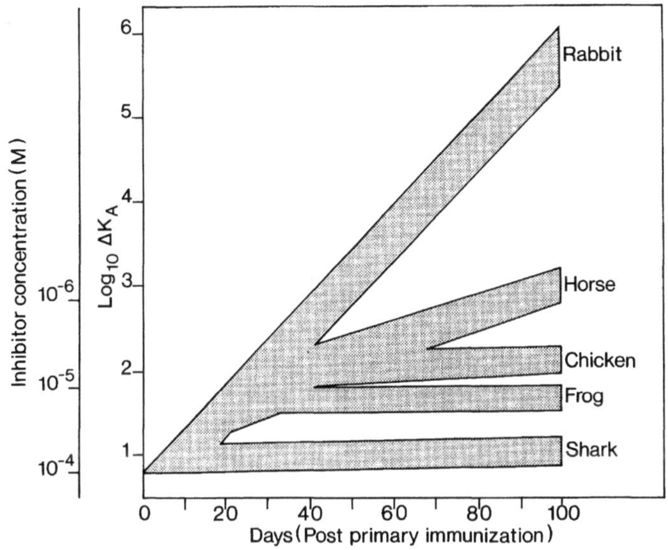

Fig. 4. Affinity maturation in various vertebrate species. Rabbit, horse, chicken, and shark data were obtained by equilibrium dialysis of purified 7S antibodies (Voss and WATT 1977). K_A is the average intrinsic association constant. Frog data were obtained in *Xenopus* (similar results can be obtained in *Rana*) by measurement of relative affinity using various concentrations of inhibitors to displace interactions of modified bacteriophage (DU PASQUIER and HAIMOVICH 1976)

mutations in the *Xenopus* and horned shark Ig was not increased significantly in CDR over framework regions.

It is difficult to reconcile how germline V products can be selected in a specific primary response (as shown in well controlled studies in *Xenopus*), but the somatic mutants cannot, or not to the same extent. Selection of *Xenopus* B cells obviously occurs since specific immune responses can be engendered, but the "circumstances" are not identical in primary and secondary responses, nor are they the same at all stages of development. In *Xenopus* tadpoles of 8–12 days of age, B cells numbers are so small that each V product is only represented in a very small number of cells, if not one cell alone, as shown by PCR-generated libraries from such animals (MUSSMANN and DU PASQUIER, unpublished data). This repertoire is apparently as heterogenous and "useful" as possible at this early stage. When immune competence emerges a few days later with the differentiation of the spleen, sequence differences among the various V_H genes, based on family specificity and unique rearrangements, are sufficient to allow selection by an immunogen. Later in the course of antigen-specific responses, after cell numbers have increased, affinity differences between germline V_Ho and the somatically-generated V_Ho^{+1} mutation might be too low and the cell dilution factor too high to allow selection in what can be considered a sub-optimal microenvironment, i.e., the amphibian lymphoid tissues lacking germinal centers and follicular dendritic cells (FDC). In summary we suggest that during the development of the immune system and during the initiation of an immune response what matters most is not whether the given B cell specificity has been generated by mutation or a rearrangement event, but whether an antigen-specific B cell is one among ten cells or 10^5 cells.

If the premise that a lack of a selecting environment inhibits the recruitment of somatic mutants, why do Ig genes in *Xenopus* and horned shark (and by extension, all ectotherms) B cells mutate at all in the course of an antigen-specific response? One possibility is that the mutational mechanism originally arose in evolution to modify Ig genes that did not rearrange (Du Pasquier 1982; Wagner and Neuberger 1996), i.e., the primary selection may have been one similar to the situation in sheep where a single V_L can be modified by mutation to generate the primary repertoire (Reynaud et al. 1991). Thus, in mature frog and shark B cells undergoing an antigen-specific response, the mutational mechanism may be expressed "neotenically" (in an evolutionary sense), perhaps in a similar fashion to re-expression of RAG genes in mouse germinal center B cells (Han et al. 1996). It is possible that generation of mutants, in the absence of an efficient selection mechanism, may still provide animals with an expanded and potentially useful repertoire.

If germinal centers indeed play such a key role in selection, then mammals made germinal center-deficient should lack affinity maturation. Two such models have arisen by chance when lymphotoxin (LT) a or CD19 "knock-out mice" (KO) were generated, and these two mutants have yielded contradictory results. In the LTa KO (Matsumoto et al. 1996) it was suggested that antibody affinity maturation can take place without germinal centers: the immunized mice had the same somatic mutations that are typically detected in responses of wild-type mice, and R/S ratios seemed, according to the authors, to indicate a selection of mutants in CDR. However, given the number of mutants analyzed in these mice, the R/S ratio is not significantly different from that expected without selection (7.7 vs 4.9, $p > 0.3$). Moreover, affinity maturation was estimated neither by equilibrium dialysis nor a stepwise titration of an inhibition of the antigen-antibody reaction, but rather an imprecise estimation of low or high affinity was based on a 4-fold level of hapten substitution of the test antigen only. In other words, the germinal center-deficient LTa KO mouse may have indeed recapitulated the situation in ectotherms. In contrast to the LTa KO mice, the CD19 KO also lacked germinal centers but showed no evidence of affinity maturation (Rickert et al. 1995).

It is not well appreciated but even among mammals dramatic differences exist in the final affinity of specific antibody in rabbits, chicken, goats, and horses (Voss and Watt 1977). For some antigens, as great as a one thousand-fold difference in affinity can be measured at the peak of affinity maturation (see Fig. 4). It is possible, therefore, that either the mechanism of somatic mutation or the way that mutants are selected can vary even at this late stage of vertebrate evolution. This may not be surprising given that rabbit, chicken, and sheep build up their pre-immune repertoire differently from mouse and human (reviewed in Du Pasquier 1993); thus, lymphoid organ complexity may not be identical among endotherms and profound differences might occur at the level of selection.

3.1 Different Levels of Selection: The Case of NAR

Perhaps, in addition to their contribution to antibody diversity, somatic variants are selected or counter-selected on other structural bases. It is well established that the R/S ratio is very low in framework regions (e.g., almost always 1.6 in FR1; SHLOMCHIK et al. 1990) because mutations within this region can disrupt the structure of the Ig domain. NAR seems to be another example of how mutants are generated and selected on structural considerations.

Because the NAR gene can have such a high frequency of mutations, we propose that it may serve an important function for the shark's immune response. It is possible that IgM, being present in such large amounts in serum and seemingly with very low frequencies of mutation, may be the first line of defense in an immune response and NAR (and perhaps IgNARC, another secreted antigen receptor in sharks related to NAR; GREENBERG et al. 1996), provides the shark with more efficient specific immunity. However, the R/S ratios of the NAR CDR1 and 2 do not indicate that the cDNAs analyzed are under stringent antigenic selection (see Figs. 2, 3). Thus, even when the frequency of mutations is high in antigen-reactive cells of ectotherms, in the absence of germinal centers such mutant cells may not be optimally selected.

Some NAR residues (e.g., Ser 33 in Fig. 2, Ser 41 in Fig. 5), however, seem to be under a strong positive selection based on the R/S values, and some of these amino acids should be in position to interact with antigen. Furthermore, in many cases synonymous substitutions are found in codons that have also mutated non-synonymous sites (see Figs. 2, 3, 5). Electron microscopic evidence strongly suggests that the V regions of NAR interact with antigen as single entities and not as dimers found for bona fide Ig or TCR a/b and g/d (ROUX and FLAJNIK, unpublished data). Thus, different regions of the NAR V domain can be exposed to the solvent, perhaps in a fashion similar to that recently described for camel IgG (DESMYTER et al. 1996). This is exemplified by the relatively large number of mutations in FR2 (see Fig. 5). Conversely, there are structural constraints on an unassociated, single V not found in bona fide Ig. For example, we believe that the non-canonical cysteine in FR2 (psn 35 in Fig. 5) is likely to be involved in forming a disulfide bond with another non-canonical cysteine encoded in CDR3; indeed, this FR2 cysteine is under strong selection against change (1R/8S), consistent with our hypothesis. In the same vein, many mutations that could have affected the joining with L chains might be neutral in NAR and would appear as being counter-selected, e.g., residue Lys-40 is a phylogenetically conserved Glu in almost all Ig/TCR V domains and is not only different in NAR, but its codon is prone to mutation.

FR 2	34	35	36	37	38	39	40	41	42	43	
Residue	T	C	W	Y	R	K	K	S	G	S	
germline	ACG	TGC	TGG	TAT	CGA	AAA	AAA	TCG	GGC	TCA	
1	---	---	---	---	---	---	---	---	---	---	
2	---	---	---	---	---	---	---	---	---	---	
3	---	---	---	---	---	---	---	---	---	---	
4	---	---	---	---	---	---	---	C--	---	---	1/0
5	---	---	---	-T-	---	---	---	---	---	---	1/0
6	---	---	---	---	---	---	---	-T-	---	---	1/0
7	---	---	---	---	---	---	---	---	---	---	
8	---	--T	---	G--	---	---	---	---	---	AGC	3/2
9	---	---	---	GG-	---	CCT	GTG	C--	-A-	AG-	11/1
10	---	---	---	---	---	---	---	---	---	---	
11	---	--T	---	---	---	---	-GC	C--	---	---	3/1
12	---	---	---	---	---	---	---	---	---	---	
13	--C	---	---	---	---	---	---	---	---	---	0/1
14	C--	---	---	---	---	-C-	---	---	---	---	2/0
15	---	---	---	---	---	---	---	---	---	---	
16	---	---	---	---	---	C--	--G	G--	-A-	---	3/1
17	---	---	---	---	--C	-GC	---	-T-	--G	---	3/2
18	---	--T	---	---	---	---	--G	G--	---	---	1/2
19	---	---	---	-T-	---	---	---	---	---	---	1/1
20	---	--T	---	---	---	---	---	---	---	---	0/1
21	---	---	---	---	--G	-T-	-G-	---	---	---	2/1
22	---	--T	---	---	---	---	---	---	---	---	0/1
23	---	---	---	---	---	-CG	CT-	CA-	---	---	5/1
24	-AC	---	---	---	---	--G	---	C--	---	GG-	4/2
25	---	--T	---	---	---	--G	TT-	---	---	CG-	4/2
26	---	---	---	---	---	---	---	---	---	---	
27	---	--T	---	---	--G	-CG	---	---	---	--C	1/4
28	---	---	---	---	---	---	---	---	---	---	
29	---	-TT	---	---	---	---	---	---	---	---	1/1
30	---	---	---	CT-	--G	---	---	---	---	---	2/1
31	---	---	---	---	---	---	CC-	---	---	--T	2/1
	2R	1R		7R	0R	9R	12R	10R	2R	8R	
	2S	8S		0S	4S	4S	4S	0S	1S	3S	

Fig. 5. Nucleotide sequence of NAR framework 2 (FR2) in cDNA clones may reveal a different type of selection based on structure. The number of replacement (R) and silent (S) substitutions are indicated for each codon and each FR2. Note the selection against change in the codon encoding the non-canonical cysteine residue 35. Total residues, 10; R/S ratio, 51:26 = 1.96

4 Molecular Nature of the Mutations and Patterns of Base Substitutions Among Vertebrates: Where a GC Bias?

Given the possible important differences in selection of mutants among vertebrates, we wondered whether those differences were reflected in the pattern of base substitutions. Most mutations in bona fide Ig in amphibians and chondrichthyans involved single base substitutions. In *Xenopus* two of the 56 mutations involved deletions of a few base pairs. Such a deletion also occurred in NAR.

The survey of *Xenopus* and horned shark Ig mutations reveals a striking preference for alteration of GC rather than AT bases (Tables 1–3), and also reflects a preference for transitions and no strand bias. The GC bias is also apparent in the proposed variants of the expressed germline joined *Raja* genes (ten GC bases mutated vs two AT) which not only reinforces the general occurrence of this bias in lower vertebrate Ig, but also suggests that those mutants are somatic. The GC bias

Table 1. Horned shark pattern of base substitutions. [From HINDS et al. (1993)]

To	From A	G	C	T	Total
A	–	16	5	1	22
G	2	–	1	1	4
C	4	6	–	2	12
T	3	3	10	–	16
Total	9	25	25	4	54

Table 2. *Xenopus* pattern of base substitutions. [From WILSON et al. (1992a, b, 1995)]

To	From A	G	C	T	Total
A	–	18	8	0	26
G	2	–	7	0	9
C	0	6	0	2	8
T	0	1	12	–	13
Total	2	25	27	2	56

Table 3. NAR pattern of base substitutions.[a] [From GREENBERG et al. (1995)]

To	From A	G	C	T	Total
A	–	48 (45)	28 (35)	19 (23)	95 (103)
G	68 (53)	–	46 (57)	31 (57)	145 (147)
C	58 (45)	51 (47)	–	35 (42)	144 (134)
T	56 (44)	24 (22)	59 (73)	–	139 (139)
Total	182 (142)	123 (114)	132 (165)	85 (102)	523 (523)

[a]Numbers in parentheses in each box are values normalized to the relative percentages of each base in the NAR V region (A, 32%; G, 27%; C, 20%; T, 21%).

is not found as a rule in mammals (see below) suggesting that in Ig genes of cold blooded vertebrates the somatic mutation machinery is somehow incomplete and affects bases asymmetrically. This would be consistent with the results of BACHL and WABL (1996) suggesting that the mechanisms of mutation can be manifold; one of the mechanisms revealed in the 18.81 mouse lymphocyte cell line also affects GC base pairs. Furthermore, BETZ et al. (1993), classified the mutational hotspots in mouse Ig genes into selected (because they seem to have been selected by antigens in CDR) and intrinsic. No GC bias was observed in the selected hotspots, whereas in the unselected ones 111 out of 123 mutations affected a G or a C.

Thus, the GC bias may not have been seen previously in mammals because it is overcome either by selection or by a qualitative change in the mutational machinery. The oxidized base 8-hydroxydeoxyguanine is found at relatively high levels in mammalian DNA (Richter et al. 1988) and causes misreading of DNA templates (Kuchino et al. 1987). This simple mechanism, therefore, may account for the GC preference in the primary, unselected mutations. This "selection hypothesis" could be tested in several other ways:

1. The mutations in germinal center-deficient mice should indicate whether the high GC/AT mutation ratio is a common finding. We analyzed the LTa KO vs wildtype mice and found that 65 out of 110 mutations (59%) affected a G or a C in wild-type, whereas in this KO mouse the ratio was 32/41 (72%) and $p < 0.05$. Thus, consistent with our interpretation, there is a significant trend towards a GC bias in this situation.
2. In non-Ig genes a tabulation of the spontaneous mutations in systems involving *aprt* and *bprt* gene inactivation revealed that the logarithm of the GC/AT mutation ratio is positively correlated with the percentage of point mutations and hence negatively correlated with the stringency of selection (Wilson et al. 1992b).

These data are all compatible with the following notions: (1) Hypermutation at Ig loci is caused by an impedance of the error correction machinery present in all cells rather than by the presence of a novel error-prone repair system in B cells; (2) antigenic selection of B cells may overcome the inherent GC bias in the mutational mechanism and thus selection in *Xenopus* is rather primitive.

Inconsistent with the "selection" hypothesis is the finding that a high GC/AT mutation ratio is not found in non-coding regions of mouse Ig genes (Lebecque and Gearhart 1990) or in a hybridoma involving a mouse myeloma and a pre-B cell line (Green et al. 1995). Nor is the GC bias detected in the NAR mutants where the incidence of mutation is thus not only qualitatively but also quantitatively different (Tables 1–3). Furthermore, a set of non-Ig sequences (human b globin and the prokaryotic genes *neo* and *gph*) submitted to somatic mutations after their insertion into the Ig locus, shows different patterns for each substrate, even though all of these genes contain the mutational hotspots (Yelamos et al. 1996). We can conclude that the GC bias is not universal and may be specific of cold blooded vertebrate Ig genes, or unsuspected selection and/or mechanisms are at work in different systems. It is not inconceivable that, if hypermutation mechanisms employ up- or down-regulation of enzymes involved in mismatch repair (e.g., those described in Gallinari and Jiricny 1996), perhaps different ones are affected in different species or even among different gene families within species.

5 Relationships Between the Various Mechanisms Generating Diversity Somatically

Table 4 summarizes diversity-generating mechanisms in vertebrates and could perhaps help in deciding what might have been the evolutionary pressures to

Table 4. The contribution of the various somatic mechanisms to the diversification of immunoglobulin genes (H locus)

Mechanism		CD Regions affected			Vertebrate classes where the mechanism is found
		CDR1	CDR2	CDR3	
Rearrangement, junctional diversity (P diversity)		–	–	+	All gnathostomes
TdT (N-diversity)		–	–	+	All gnathostomes?
Multiple D usage		–	–	+	All gnathostomes
Multiple J usage		–	–	+	All gnathostomes except birds
Multiple V usage		+	+	+	All gnathostomes except birds and rabbits
Somatic mutation	post rearrangement	+	+	+	Chondrichthyans, amphibians, birds and Mammals
Gene conversion	post rearrangement	+	+	+	Birds, mammals
Secondary rearrangement	post rearrangement	+	+	+	Amphibians, mammals

develop or maintain a hypermutation mechanism in the immune system. During V(D)J rearrangement no mechanism other than combinatorial joining can create diversity in CDR1 and CDR2. After rearrangement three mechanisms, so far represented only in jawed vertebrates, can affect the specificity of the antibody molecule by modifying CDR1 and CDR2. Gene conversion on which we shall not expand is clearly an adaptation to a system lacking combinatorial joining because of the paucity of utilized germline V genes (birds and rabbits). Somatic mutations resulting in single amino acid modifications or deletions/insertions refine an existing specificity without a complete alteration of the CDRs. The third mode of post-rearrangement modification of the CDR is V gene replacement due to secondary rearrangement permitted by the structure of the L chain locus, (receptor editing; Tiegs et al. 1993), and by the possibility of bypassing D element, because of the presence of cryptic RSS in V_H FR3 (Schwager et al. 1991; Chen et al. 1994; Radic and Zouali 1996). If one ignores the apparent lack of usefulness of such a mechanism in species that show ongoing rearrangements from the complete genome during all their life, one can speculate on the evolutionary selection for this mechanism, especially since it has been observed in amphibians (Fig. 6).

As originally proposed in mammals (Shlomchik et al. 1990; Chen et al. 1994), V gene replacement can generate major changes in specificity that may result in escape from autoimmunity without losing previously acquired characteristics of the cell; that is to say, useful V_H rearrangements are maintained while the V_L is exchanged. Thus, if the biological role of such a mechanism is very different from that of somatic mutation, one could understand its significance. For example, in species with low cell numbers and/or having only one wave of rearrangement, this mechanism could preserve lymphocytes as they escape autoantigens. This situation could easily be encountered in some ectotherms such as amphibians at metamorphosis or young fish and amphibian larvae with very few lymphocytes. In addition, the evolutionary selecting force may have been, as described for developing T cells (Borgulya et al. 1992), a lack of positive selection rather than an escape from self censorship.

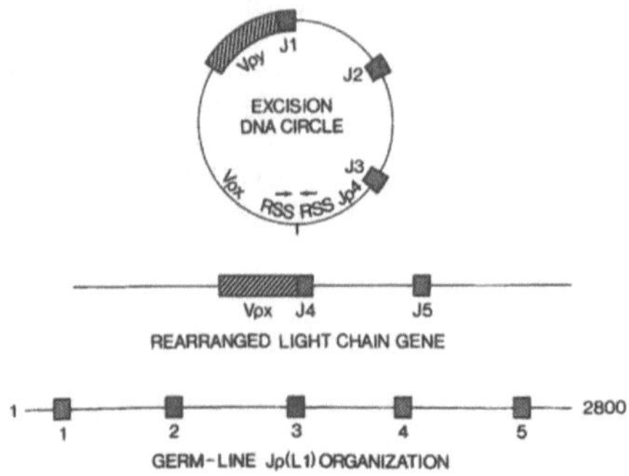


```
                          octamer
  1 TGTAACACAGAAATGTATCTGTTGATTTGCATACACAAAGTGTGTGGTGAGTTCCCTGTAGATATAAGATGCTATAAATGTTCTATTTCCTATGTGACGT
    ----+----1----+----2----+----3----+----4----+----5----+----6----+----7----+----8----+----9----+----0

                                                                     leader 1
                                               MetGlyArgTyrLeuLeuLeuValLeuSerThrLeuPheSerLeuSerValPro
101 CTCAGTGATCTGCACCTGACACTGCTCCTACTGTCAGTATGGGGAGATACTTACTGCTTGTCCTCTCCACTCTCTTCTCTCTCTGTTCCAGGTAATTA
    ----+----1----+----2----+----3----+----4----+----5----+----6----+----7----+----8----+----9----+----0

201 TACCAACACTCACTTACATTGATTTCCCTAGTGATTTTGAGCTATATAGAACTTACTTTTGGATTTGTAAATATCTGAATAAAATGACTTTATTTTTGAA
    ----+----1----+----2----+----3----+----4----+----5----+----6----+----7----+----8----+----9----+----0

                 leader 2         I1
                 GlyPheTyrGlyGlnIleValLeuThrGlnSerProAspTyrValSerValSerProGlyGluThrValThrLeuThrCysLys
301 TTTTCTAATTTTTTCAGGTTTCTATGGGCAGATTGTACTGACTCAGTCTCCAGATTATGTCTCTGTGTCACCAGGAGAAACTGTCACCCTCACATGTAAA
    ----+----1----+----2----+----3----+----4----+----5----+----6----+----7----+----8----+----9----+----0

                                                                             V rau
    AlaSerSerSerValTyrAsnAlaIleAlaTrpTyrGlnGlnLysSerGlyGlnAlaProLysLeuLeuIleTyrGluAlaAsnThrArgHisThrGlyT
401 GCCAGTAGCAGTGTTTATAATGCCATAGCCTGGTACCAACAGAAATCAGGGCAGGCACCAAAGCTGCTGATCTATGAGGCAAACACCCGACACACAGGGA
    ----+----1----+----2----+----3----+----4----+----5----+----6----+----7----+----8----+----9----+----0

    hrProGluArgIleSerAlaSerArgSerGlyAsnAlaPheThrLeuThrIleSerArgMetGluAlaGluAspAlaAlaAspTyrTyrCysGlnGlnGl
501 CCCCAGAGCGGATCAGCGCCAGTAGGTCTGGAAATGCTTTCACTCTTACAATCAGCAGAATGGAAGCAGAAGATGCAGCAGATTATTACTGTCAGCAGGG
    ----+----1----+----2----+----3----+----4----+----5----+----6----+----7----+----8----+----9----+----0

              I        J  rau 4                                         intron j rau 4 - j rau 5
    yTyrGlySerProTyrThrPheGlyLysGlyThrArgValGluIleLys
601 TTACGGTTCCCCGTACACTTTTGGTAAAGGAACCAGGGTAGAGATAAAACGTAATATTCCTTATAAGACTTATTTATCGCTGTTATCACTTTTTCAGTAA
    ----+----1----+----2----+----3----+----4----+----5----+----6----+----7----+----8----+----9----+----0

701 ATGTAATGGTAGTTTGATTTTGAATTGTTTATATACTTTT
    ----+----1----+----2----+----3----+----4
```

Fig. 6. Example of a secondary rearrangement in the light chain locus rau of *Xenopus*. Graphic interpretation and actual sequence of the Vrau-Jrau4 segment obtained from a spleen/thymus circular DNA library. The library was screened with a synthetic oligonucleotide corresponding to a head–head join between the RSS

6 Introduction of Somatic Mutations Into the Immune System

Clearly the experiments described in this chapter negate the hypothesis that the mutational process does not exist in cold blooded vertebrates. In fact, mutation may even have preceded the rearrangement mechanism; a priori, a repertoire generated by either of these processes could provide lymphocytes with a large, selectable repertoire. The somatic hypermutation mechanism in B cells is targeted, with a high rate of mutations that are selectable. Each of these three features – target, high rate, and selection – have evolutionary origins that have not been elucidated. In the following scenarios, we assume that the somatic hypermutation mechanism in lymphocytes did not arise de novo, but rather is derived from existing somatic mutation "mechanisms".

Scenario 1: The Ability to Select Somatic Mutants Arose First. At this early stage, we assume mutation affects the whole genome (no targeting yet), but the rate of mutation was not high enough in the whole body to jeopardize survival. Then, with a poor rate, in order to select some useful mutants, a large number of cells must exist in the individual. This may not have been the case in immunocompetent animals with very few lymphocytes. From previous discussion, obviously we believe that stringent selection mechanisms arose late in vertebrate evolution.

Scenario 2: The Rate of Somatic Mutations Increased First in the Individual and Before Targeting. Here, the individual can survive mutations only if not all cells run the risk of producing a deleterious mutant; this would be possible only in species with a relatively small genome size. It has been argued that somatic mutation favored the evolution of diploidy (Orr 1995); perhaps tetraploidy has favored its introduction into the immune system. If the rate of mutation is 1×10^{-6}/bp per generation in an individual with $1'10^{9}$/bp per haploid genome, each individual will carry 1000 mutants. If the genome size is reduced, the number of mutation will also be reduced. An increase in the overall rate of mutation before targeting could indeed be tolerated without too much danger if the constitution of a large genome is due to polyploidy or the presence of repetitive sequences rather than to an increase in the number of useful and necessary genes. Under these two circumstances (presence of several independent copies of one gene, or "dilution" of the gene in a large amount of non-coding sequence) the increase in rate would have little chance to be lethal, and could have occurred before targeting. That vertebrates have increased their genome size by polyploidization is well documented (see Lundin 1993; Kasahara et al. 1997). With some rudimentary targeting, i.e., into immune cells only, cell wastage could have been accompanied by an increase in diversity in antigen receptors.

Scenario 3: The Targeting Occurred First. This could have been a security. For this, there were perhaps two steps: targeting into a lymphocyte lineage and then into a specific area of the genome. But in the absence of mechanisms to select mutants or generate a high rate of mutation, there was perhaps not much to target!

However, if the targeted mutational mechanism indeed arose in order to generate the preimmune repertoire, perhaps in species in which diversity was not generated through rearrangement, the only selection required would have been for a properly folded antigen receptor. We look forward to the next wave of research into the evolution of somatic mutation, when ectotherms may provide models to further address controversies surrounding mechanisms and origins. It will be especially interesting to determine whether targeted mutational processes might occur in animals, like the jawless vertebrates and protochordates, in which a bona fide adaptive immune system is yet to be discovered.

Acknowledgements. We thank Pat Washington for preparation of the manuscript and Marilyn Diaz and M.C. Steinberg for discussions of the models. The work described herein on NAR was supported by National Institutes of Health grant RR06603. The Basel Institute for Immunology was founded and is supported by Hoffmann La Roche, Ltd.

References

Anderson M, Amemiya C, Luer C, Litman R, Rast J, Nimura Y, Litman G (1994) Complete genomic sequence and patterns of transcription of a member of an unusual family of closely related, chromosomally dispersed Ig gene clusters in Raja. Int Immunol 6:1661–1670

Anderson MK, Shamblott MJ, Litman RT, Litman GW (1995) Generation of immunoglobulin light chain gene diversity in Raja erinacea is not associated with somatic rearrangement, an exception to a central paradigm of B cell immunity. J Exp Med 182:109–119

Bachl J, Wabl M (1996) An immunoglobulin mutator that targets G.C base pairs. Proc Natl Acad Sci USA 93:851–855

Betz AG, Neuberger MS, Milstein C (1993) Discriminating intrinsic and anti-selected mutational hotspots in immunoglobulin V genes. Immunol Today 14:405–411

Brandt DC, Griessen M, Du Pasquier L, Jaton JC (1980) Antibody diversity in amphibians: evidence for inheritance of idiotypic specificities in isogeneic Xenopus. Eur J Immunol 10:731–736

Borgulya P, Kishi H, Uematsu Y, von Boehmer H. (1992) Exclusion and inclusion of a and b T cell receptor alleles. Cell 69:529–537

Chen C, Radic MZ, Erikson J, Camper SA, Litwin S, Hardy RR, Weigert M (1994) Deletion and editing of B cells that express antibodies to DNA. J Immunol 152:1970–1982

Clem LW, Leslie GA (1971) Production of 19S IgM antibodies with restricted heterogeneity from sharks. Proc Natl Acad Sci USA 68:139–143

Desmyter A, Transue TR, Ghahrondi MA, Thi M-HD, Poortmans F, Hamers R, Muyldermans S, Wyns L (1996) Crystal structure of a camel single-domain V_H antibody fragment in complex with lysozyme. Nature Struct Biol 3:803–811

Du Pasquier L (1982) Antibody diversity in lower vertebrates – why is it so restricted? Nature 276:311–313

Du Pasquier L (1993) Phylogeny of B-cell development. Curr Opin Immunol 5:185–193

Du Pasquier L, Haimovich J (1976) The antibody response during amphibian ontogeny. Immunogenetics 3:381–391

Du Pasquier L, Robert J (1992) In vitro growth of thymic tumor cell lines from Xenopus. Dev Immunol 2:295–307

Gallinari P, Jiricny J (1996) A new class of uracil-DNA glycosylases related to human thymine-DNA glycosylase. Nature 383:735–738

Green NS, Rabinowitz JL, Zhu M, Kobrin BJ, Scharff MD (1995) Immunoglobulin variable region hypermutation in hybrids derived from a pre-B- and a myeloma cell line. Proc Natl Acad Sci USA 92:6304–6308

Greenberg AS, Avila D, Hughes M, Hughes A, McKinney EC, Flajnik MF (1995) A new antigen receptor gene family that undergoes rearrangement and extensive somatic diversification in sharks. Nature 374:168–173

Greenberg AS, Hughes AL, Guo J, Avila D, McKinney EC, Flajnik MF (1996) A novel "chimeric" antibody class in cartilaginous fish: IgM may not be the primordial immunoglobulin. Eur J Immunol 26:1112–1129

Han S, Zheng B, Schatz DG, Spanopoulou E, Kelsoe G (1996) Neoteny in lymphocytes: Rag1 and Rag2 expression in germinal center B cells. Science 274:2094–2097

Hinds KR, Litman GW (1986) Major reorganization of immunoglobulin V_H segmental elements during vertebrate evolution. Nature 320:546–549

Hinds-Frey KR, Nishikata H, Litman RT, Litman GW (1993) Somatic variation precedes extensive diversification of germline sequences and combinatorial joining in the evolution of immunoglobulin heavy chain diversity. J Exp Med 178:815–824

Kasahara M, Nakaya J, Satta Y, Takahata N (1997) Chromosomal duplication and the emergence of the adaptive immune system. Trends Genet 13:90–92

Kokubu F, Hinds K, Litman R, Shamblott MJ, Litman GW (1987) Extensive families of constant region genes in a phylogenetically primitive vertebrate indicate an additional level of immunoglobulin complexity. Proc Natl Acad Sci USA 84:5868–5872

Kokubu F, Hinds K, Litman R, Shamblott MJ, Litman GW (1988) Complete structure and organization of immunoglobulin heavy chain constant region genes in a phylogenetically primitive vertebrate. EMBO J 7:1979–1988

Kuchino Y, Mori F, Kasai H, Inoue H, Iwai S, Miura K, Ohtsuka E, Nishimura S (1987) Misreading of DNA templates containing 8-hydroxydeoxyguanosine at the modified base and at adjacent residues. Nature 327:77–79

Lebecque SG, Gearhart PJ (1990) Boundaries of somatic mutation in rearranged immunoglobulin genes: 5′ boundary is near the promoter, and 3′ boundary is ~1 kb from V(D)J gene. J Exp Med 172:1717–1727

Lundin LG (1993) Evolution of the vertebrate genome as reflected in paralogous chromosomal regions in man and the house mouse. Genomics 16:1–19

MacLennan IC (1994) Somatic mutation: from the dark zone to the light. Curr Biol 4:70–72

Mäkelä O, Litman GW (1980) Lack of heterogeneity in anti-hapten antibodies of a phylogenetically primitive shark. Nature 287:639–641

Matsunaga T (1985) Evolution of antibody repertoire-somatic mutation as a latecomer. Dev Comp Immunol 9:585–596

Matsumoto M, Lo SF, Carruthers CJ, Min J, Mariathasan S, Huang G, Plas DR, Martin SM, Geha RS, Nahm MH, Chaplin DD (1996) Affinity maturation without germinal centers in lymphotoxin-alpha-deficient mice. Nature 382:462–466

Mussmann R, Wilson M, Marcuz A, Courtet M, Du Pasquier L (1996) Membrane exon sequences of the three Xenopus Ig classes explain the evolutionary origin of mammalian isotypes. Eur J Immunol 26:409–414

Orr HA (1995) Somatic mutation favors the evolution of diploidy. Genetics 139:1441–1447

Radic MZ, Zouali M (1996) Receptor editing, immune diversification, and self tolerance. Immunity 5:505–511

Reynaud C-A, Mackay CR, Müller RG, Weill J-C (1991a) Somatic generation of diversity in a mammalian primary lymphoid organ: the sheep ileal Peyer's patches. Cell 64:995–1005

Richter C, Park J-W, Ames BN (1988) Normal oxidative damage to mitochondrial and nuclear DNA is extensive. Proc Natl Acad Sci USA 85:6465–6467

Rickert RC, Rajewsky K, Roes J (1995) Impairment of T-cell-dependent B-cell responses and B-1 cell development in CD19-deficient mice. Nature 376:352–355

Schwager J, Bürckert N, Courtet M, Du Pasquier L (1989) Genetic basis for the antibody repertoire in Xenopus. Analysis of the V_H diversity. EMBO J 8:2989–3001

Schwager J, Bürckert N, Courtet M, Du Pasquier L (1991) The ontogeny of diversification at the immunoglobulin heavy chain locus in Xenopus. EMBO J 10:2451–2470

Shlomchik M, Mascelli M, Shan H, Radic MZ, Pisetsky D, Marshak-Rothenstein A, Weigert M (1990) Anti DNA antibodies from autoimmune mice arise by clonal expansion and somatic mutation. J Exp Med 171:265–292

Tiegs SL, Russell DM, Nemazee D (1993) Receptor editing in self-reactive bone marrow B cells. J Exp Med 177:1009–1020

Voss EW, Watt RH (1977) Comparison of the microenvironment of chicken and rabbit antibody active site. Adv Exp Biol Med 88:391–401

Wagner SD, Neuberger MS (1996) Somatic hypermutation of immunoglobulin genes. Annu Rev Immunol 14:441–457

Wagner SD, Milstein C, Neuberger MS (1995) Codon bias targets mutation. Nature 376:732

Wilson M, Marcuz A, Courtet M, Du Pasquier L (1992a) Sequences of C mu and the V_H1 family in LG7, a clonable strain of Xenopus, homozygous for the immunoglobulin loci. Dev Immunol 3:13–24

Wilson M, Hsu E, Marcuz A, Courtet M, Du Pasquier L, Steinberg C (1992b) What limits affinity maturation of antibodies in Xenopus – the rate of somatic mutation or the ability to select mutants? EMBO J 11:4337–4347

Wilson M, Marcuz A, Du Pasquier L (1995) Somatic mutations during an immune response in Xenopus tadpoles. Dev Immunol 4:227–234

Yelamos J, Klix N, Goyenechea B, Lonzo F, Chui YL, Gonzales Fernandes A, Pannell R, Neuberger MS, Milstein C (1996) Targeting of non Ig sequences in place of the V segment by somatic hypermutation. Nature 376:225–229

Zapata AJ, Torroba M, Vicente A, Varas A, Sacedone R, Jimenez E (1995) The relevance of cell microenvironment for the appearance of lymphohaemopietic tissues in primitive vertebrates. Histol Histopathol 10:761–778

Zapata AG, Torroba M, Sacedon R, Varas A, Vicente A (1996a) Structure of the lymphoid organs of elasmobranchs. J Exp Zool 275:125–143

Zapata AG, Chiba A, Varas A (1996b) Cells and tissues of the immune system of fish. Fish Physiology 15:1–62

Subject Index

Current Topics in Microbiology and Immunology

Volumes published since 1989 (and still available)

Vol. 189: **Oldstone, Michael B. A. (Ed.):** Cytotoxic T-Lymphocytes in Human Viral and Malaria Infections. 1994. 37 figs. IX, 210 pp. ISBN 3-540-57259-7

Vol. 190: **Koprowski, Hilary; Lipkin, W. Ian (Eds.):** Borna Disease. 1995. 33 figs. IX, 134 pp. ISBN 3-540-57388-7

Vol. 191: **ter Meulen, Volker; Billeter, Martin A. (Eds.):** Measles Virus. 1995. 23 figs. IX, 196 pp. ISBN 3-540-57389-5

Vol. 192: **Dangl, Jeffrey L. (Ed.):** Bacterial Pathogenesis of Plants and Animals. 1994. 41 figs. IX, 343 pp. ISBN 3-540-57391-7

Vol. 193: **Chen, Irvin S. Y.; Koprowski, Hilary; Srinivasan, Alagarsamy; Vogt, Peter K. (Eds.):** Transacting Functions of Human Retroviruses. 1995. 49 figs. IX, 240 pp. ISBN 3-540-57901-X

Vol. 194: **Potter, Michael; Melchers, Fritz (Eds.):** Mechanisms in B-cell Neoplasia. 1995. 152 figs. XXV, 458 pp. ISBN 3-540-58447-1

Vol. 195: **Montecucco, Cesare (Ed.):** Clostridial Neurotoxins. 1995. 28 figs. XI., 278 pp. ISBN 3-540-58452-8

Vol. 196: **Koprowski, Hilary; Maeda, Hiroshi (Eds.):** The Role of Nitric Oxide in Physiology and Pathophysiology. 1995. 21 figs. IX, 90 pp. ISBN 3-540-58214-2

Vol. 197: **Meyer, Peter (Ed.):** Gene Silencing in Higher Plants and Related Phenomena in Other Eukaryotes. 1995. 17 figs. IX, 232 pp. ISBN 3-540-58236-3

Vol. 198: **Griffiths, Gillian M.; Tschopp, Jürg (Eds.):** Pathways for Cytolysis. 1995. 45 figs. IX, 224 pp. ISBN 3-540-58725-X

Vol. 199/I: **Doerfler, Walter; Böhm, Petra (Eds.):** The Molecular Repertoire of Adenoviruses I. 1995. 51 figs. XIII, 280 pp. ISBN 3-540-58828-0

Vol. 199/II: **Doerfler, Walter; Böhm, Petra (Eds.):** The Molecular Repertoire of Adenoviruses II. 1995. 36 figs. XIII, 278 pp. ISBN 3-540-58829-9

Vol. 199/III: **Doerfler, Walter; Böhm, Petra (Eds.):** The Molecular Repertoire of Adenoviruses III. 1995. 51 figs. XIII, 310 pp. ISBN 3-540-58987-2

Vol. 200: **Kroemer, Guido; Martinez-A., Carlos (Eds.):** Apoptosis in Immunology. 1995. 14 figs. XI, 242 pp. ISBN 3-540-58756-X

Vol. 201: **Kosco-Vilbois, Marie H. (Ed.):** An Antigen Depository of the Immune System: Follicular Dendritic Cells. 1995. 39 figs. IX, 209 pp. ISBN 3-540-59013-7

Vol. 202: **Oldstone, Michael B. A.; Vitković, Ljubiša (Eds.):** HIV and Dementia. 1995. 40 figs. XIII, 279 pp. ISBN 3-540-59117-6

Vol. 203: **Sarnow, Peter (Ed.):** Cap-Independent Translation. 1995. 31 figs. XI, 183 pp. ISBN 3-540-59121-4

Vol. 204: **Saedler, Heinz; Gierl, Alfons (Eds.):** Transposable Elements. 1995. 42 figs. IX, 234 pp. ISBN 3-540-59342-X

Vol. 205: **Littman, Dan R. (Ed.):** The CD4 Molecule. 1995. 29 figs. XIII, 182 pp. ISBN 3-540-59344-6

Vol. 206: **Chisari, Francis V.; Oldstone, Michael B. A. (Eds.):** Transgenic Models of Human Viral and Immunological Disease. 1995. 53 figs. XI, 345 pp. ISBN 3-540-59341-1

Vol. 207: **Prusiner, Stanley B. (Ed.):** Prions Prions Prions. 1995. 42 figs. VII, 163 pp. ISBN 3-540-59343-8

Vol. 208: **Farnham, Peggy J. (Ed.):** Transcriptional Control of Cell Growth. 1995. 17 figs. IX, 141 pp. ISBN 3-540-60113-9

Vol. 209: **Miller, Virginia L. (Ed.):** Bacterial Invasiveness. 1996. 16 figs. IX, 115 pp. ISBN 3-540-60065-5

Vol. 210: **Potter, Michael; Rose, Noel R. (Eds.):** Immunology of Silicones. 1996. 136 figs. XX, 430 pp. ISBN 3-540-60272-0

Vol. 211: **Wolff, Linda; Perkins, Archibald S. (Eds.):** Molecular Aspects of Myeloid Stem Cell Development. 1996. 98 figs. XIV, 298 pp. ISBN 3-540-60414-6

Vol. 212: **Vainio, Olli; Imhof, Beat A. (Eds.):** Immunology and Developmental Biology of the Chicken. 1996. 43 figs. IX, 281 pp. ISBN 3-540-60585-1

Vol. 213/I: **Günthert, Ursula; Birchmeier, Walter (Eds.):** Attempts to Understand Metastasis Formation I. 1996. 35 figs. XV, 293 pp. ISBN 3-540-60680-7

Vol. 213/II: **Günthert, Ursula; Birchmeier, Walter (Eds.):** Attempts to Understand Metastasis Formation II. 1996. 33 figs. XV, 288 pp. ISBN 3-540-60681-5

Vol. 213/III: **Günthert, Ursula; Schlag, Peter M.; Birchmeier, Walter (Eds.):** Attempts to Understand Metastasis Formation III. 1996. 14 figs. XV, 262 pp. ISBN 3-540-60682-3

Vol. 214: **Kräusslich, Hans-Georg (Ed.):** Morphogenesis and Maturation of Retroviruses. 1996. 34 figs. XI, 344 pp. ISBN 3-540-60928-8

Vol. 215: **Shinnick, Thomas M. (Ed.):** Tuberculosis. 1996. 46 figs. XI, 307 pp. ISBN 3-540-60985-7

Vol. 216: **Rietschel, Ernst Th.; Wagner, Hermann (Eds.):** Pathology of Septic Shock. 1996. 34 figs. X, 321 pp. ISBN 3-540-61026-X

Vol. 217: **Jessberger, Rolf; Lieber, Michael R. (Eds.):** Molecular Analysis of DNA Rearrangements in the Immune System. 1996. 43 figs. IX, 224 pp. ISBN 3-540-61037-5

Vol. 218: **Berns, Kenneth I.; Giraud, Catherine (Eds.):** Adeno-Associated Virus (AAV) Vectors in Gene Therapy. 1996. 38 figs. IX,173 pp. ISBN 3-540-61076-6

Vol. 219: **Gross, Uwe (Ed.):** Toxoplasma gondii. 1996. 31 figs. XI, 274 pp. ISBN 3-540-61300-5

Vol. 220: **Rauscher, Frank J. III; Vogt, Peter K. (Eds.):** Chromosomal Translocations and Oncogenic Transcription Factors. 1997. 28 figs. XI, 166 pp. ISBN 3-540-61402-8

Vol. 221: **Kastan, Michael B. (Ed.):** Genetic Instability and Tumorigenesis. 1997. 12 figs.VII, 180 pp. ISBN 3-540-61518-0

Vol. 222: **Olding, Lars B. (Ed.):** Reproductive Immunology. 1997. 17 figs. XII, 219 pp. ISBN 3-540-61888-0

Vol. 223: **Tracy, S.; Chapman, N. M.; Mahy, B. W. J. (Eds.):** The Coxsackie B Viruses. 1997. 37 figs. VIII, 336 pp. ISBN 3-540-62390-6

Vol. 224: **Potter, Michael; Melchers, Fritz (Eds.):** C-Myc in B-Cell Neoplasia. 1997. 94 figs. XII, 291 pp. ISBN 3-540-62892-4

Vol. 225: **Vogt, Peter K.; Mahan, Michael J. (Eds.):** Bacterial Infection: Close Encounters at the Host Pathogen Interface. 1998. 15 figs. IX, 169 pp. ISBN 3-540-63260-3

Vol. 226: **Koprowski, Hilary; Weiner, David B. (Eds.):** DNA Vaccination/Genetic Vaccination. 1998. 31 figs. XVIII, 198 pp. ISBN 3-540-63392-8

Vol. 227: **Vogt, Peter K.; Reed, Steven I. (Eds.):** Cyclin Dependent Kinase (CDK) Inhibitors. 1998. 15 figs. XII, 169 pp. ISBN 3-540-63429-0

Vol. 228: **Pawson, Anthony I. (Ed.):** Protein Modules in Signal Transduction. 1998. 42 figs. IX, 368 pp. ISBN 3-540-63396-0

MIX
Papier aus verantwortungsvollen Quellen
Paper from responsible sources
FSC® C105338

If you have any concerns about our products,
you can contact us on
ProductSafety@springernature.com

In case Publisher is established outside the EU,
the EU authorized representative is:
Springer Nature Customer Service Center GmbH
Europaplatz 3, 69115 Heidelberg, Germany

Printed by Libri Plureos GmbH
in Hamburg, Germany